Book One of the *Outer*

Dr Thomas (Tom) F.
Father of Market Intelligence

| TomGrooms.com | Visit Tom ...

It's like the wild-wild-west ... the race is on to outerspace. Be the first to stake your claim in the extraterrestrial.

This first book is an introduction to the extraterrestrial and the foundation for book two ... the Space Intelligence Agency. The history of outer space has been invisible to most of us.

We live in a time when discoveries of phenomena are increasing in frequency. The ongoing occurrences of contact with ETs and UFOs are drawing our attention to questions about outer space.

In the past we have searched our souls for answers of life and now we find ourselves searching outer space for answers about our past.

The journey we are about to take together will be a cosmic one. It will take us down the trails through time and contacts of things we do not understand ... for the moment.

ALSO BY THOMAS FLETCHER GROOMS

The Market Intelligence Collection

Space Intelligence Agency ...
Spies and Espionage in the Space Race

Space Nuggets ...
Winning the Space Acquisition Race through Effective Intelligence

Alien Presence ...
Declassified ET and Alien Intelligence

Spy History ...
History of Spies and Agencies

Culture of Russia ...
The Russian Way of Business

My Mission in Russia ...
Business Incubators, Tactics and Lessons

The Collapse of Russia ...
An American Businessman's Visit During the 1998 Financial Collapse

Market Intelligence ...
The Original Work

The Executive Leadership Collection

Family Business ...
How to Create a Money Legacy

How to Start a Business in 20 Days ...
Home Business – Start A New Business - How Real Money Is Made

Declassified ET and Alien Intelligence

Intergalactic ET and UFO Visitations
On Planet Earth

Case Made for the SIA Space Intelligence Agency

FROM THE MARKET INTELLIGENCE COLLECTION

US UK Canada Europe Japan India Brasil Mexico Australia

ALIEN PRESENCE
Declassified ET and Alien Intelligence

Copyright © 2021
MKI – Market Intelligence Series
ISBN: 979-8554-90401-1 Nonfiction

The written content within contains information obtained from authentic and highly regarded sources. Reasonable efforts have been made to verify reliable data and information. The author cannot assume responsibility for the validity of all materials or the consequences of their use.

Tom Grooms is an independent author. All books are proof-read and edited twice, sometimes three times before publication. However, mistakes, omissions and spelling errors may still exist due to flawed software programs.

This book is available for review with the understanding it is copyright material and no quotation from the book may be published without proper acknowledgment of the author. The right of Thomas Fletcher Grooms to be identified as the author of this work during and after this process in perpetuity for publication has been asserted in accordance with the Copyright, Designs and Patents Act of 1988.

Printed in the United States of America, except as permitted under the United States Copyright Act of 1976 and Copyright Laws of the United Kingdom, Europe, and Worldwide. No part of this publication may be reproduced, sold, or distributed in any form or by any means, or stored in a database or retrieval system for money or income without prior written permission and remuneration to the author.

Quotation of short passages for purposes of academic research, academic papers, criticism, education, or review is an exception with proper acknowledgment of the author.

All rights reserved.

FORWARD

For a long time I struggled with writing this book and especially the third topic on ET alien encounters and UFO sightings and incidents. It was not my intent to provide a lengthy timeline on extraterrestrial phenomena merely as a page filler or some broad-based history from old to new. Rather, I hope to establish a clear and agelong pattern of ET and UFO observations.

I am aware I am risking my credibility by taking a chance writing this nonfiction book to establish the foundation for the importance of a space intelligence agency to follow in the second book of this series. You may be wondering what is out there. You may be reading this for pleasure or entertainment. You may think this is all nonsense. You may walk away becoming a true believer in the existence of ETs and UFOs.

The size of the cosmos makes it plausible there are other species out there. The age of the universe is undeterminable. We humans have been around for the tiniest of tiniest fractions on a universal timeline. Whichever the case, it was difficult deciding what information to include in revealing Declassified government and military documents in hopes of it not seeming generic.

Some of you would have wanted me to expand more here on current space trends such as such as private enterprises forging new roles in space transportation like Elon Musk and SpaceX, Jeff Bezos and Blue Origin, or those entities, private and public, planning for manned missions to Mars which we will cover in the third book in this series. I could have given you just a few sightings and events to accomplish my purpose and you may have been disappointed.

Many people think nothing like this has ever happened. Everything written regarding the events is public knowledge. Reasonable efforts were made to verify reliable data and all information. It is easier to write a book of fiction with no concern for its authenticity. My extensive research over several years shares with you information obtained from authentic and highly regarded sources.

While tomorrow is a mystery and the past is a curiosity, let's find out together as it is interesting stuff to sit around and ponder.

TOPICS

Space 1 Time to Traverse the Universe 9

Space 2 Extraterrestrial 11

    The Cosmos 15

    What the Observable Universe Looks Like 17

    Existing Civilizations in the Milky Way Galaxy 18

    Sensitive Classified Intelligence 20

    Extraterrestrial Alien Forms of Communications 21

Space 3 Extraterrestrial Alien Life Forms on Earth 23

    Earth Home to Extraterrestrial Alien Life Forms 23

    Mapping Outer Space 24

    Genetic Engineering of Intergalactic Beings 25

    Preparation of Earth Beings for Outer Space Travel 26

    Close Encounters of the Third Kind 27

    Hynek System of ET and UFO Sighting Reports Classifications 28

    Vallee System of ET and UFO Sighting Reports Classifications 31

Space 4 Extraterrestrial Law and Jurisdiction 35

    Outer Space Common Ground 35

    Interplanetary Space Law 35

    Extraterrestrial Jurisprudence 36

    Extraterrestrial Contract Law 36

    Extraterrestrial Criminal Law 36

Space 5   Revealing Sensitive Declassified ET and UFO Documents . 37

Extraterrestrial Alien Encounters and UFO Sightings & Incidents .. 38

United States and UFOs 39

1947 July 2 Roswell New Mexico UFO Crash 41

4602D Air Intelligence Service Squadron 42

Project Blue Book 43

Declassified Extraterrestrial Alien Encounters and UFO Events...... 44

Space 6   Governments and Militaries Worldwide ET-UFO Encounters 45

Russia and UFOs 49

China and UFOs 54

Fast Radio Burst Radio Telescopes and Military UFO Sightings 57

Chile and UFOs 59

Italy and UFOs 65

ET UFO Military Capability 65

How to Destroy an Enemy Military ET UFO 65

Australia and UFOs 66

Space 7   Space Forces 67

Russia Space Force 67

China Space Force 68

United States Space Force 69

Japan Space Force 70

FVEY Five Eyes 71

Space 8   Science and the Unexplained Reality of ETs and UFOs 73

   Analysis Inconclusive 73

   Preponderance of Evidence 74

   Past ... Present ... Future for Outer Space 75

   Myths Symbols Facts 76

   US & Earth Military Assets and Forces at Risk from ETs & UFOs 77

   Intelligence Agencies Prepare for the Extraterrestrial 80

   Earth Is In Danger 81

   Advantage to Being First in Outer Space 83

   Outer Space Industry a Quadrillion Dollar Market 84

Space 9   Neighbors ... Moon Extraterrestrial Spy Base 85

   First Contact 89

   SETI 90

   Moon the #1 Defense Outpost 91

   Rock – Satellite - Spaceship 92

   Understanding the Moon 93

   Hollow-Moon 94

Space 10   Far Away Habitat Galaxies ... Not So Far Away 95

   80 Species of ETs on Earth 97

   UFOs in Combat a National Security Threat 98

   How to Defeat Enemy UFO Spacecraft 99

   So What Should We Do? 100

   Final Thoughts of the Author 102

Space 1
Time to Traverse the Universe

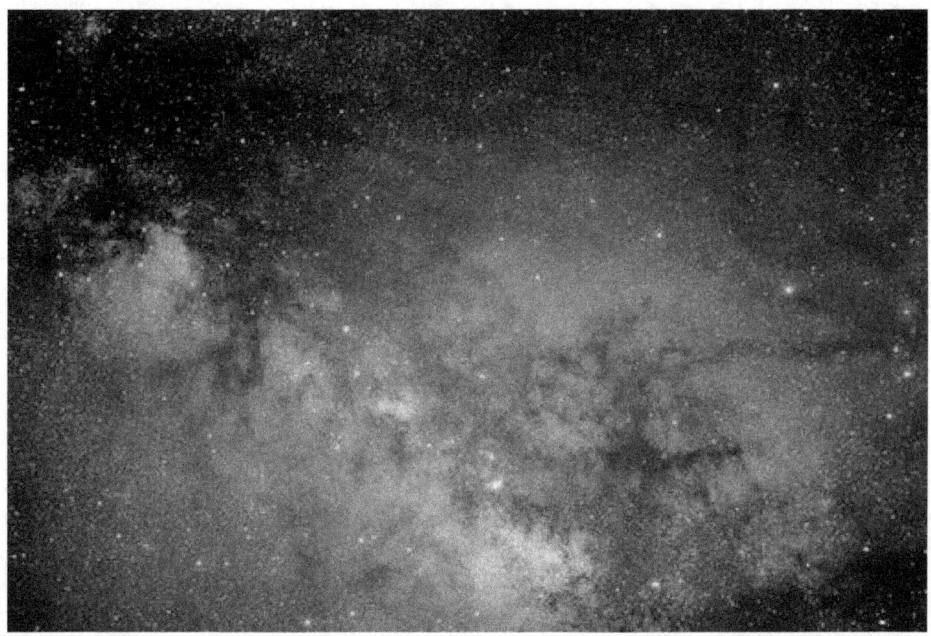

Outer space galaxies are the home for Earth's extraterrestrials.

Get hold of your mind and prepare to travel beyond anything you have ever experienced. It is here where many of your questions will be answered and your doubts are shaken.

It has come time to traverse the universe from the safety of your comfort zone. Take note and take hold of the next generation of extraterrestrials and they will come right here from Earth.

It must go without saying the nature of outer space has changed for all of us. All the quiet tranquility we have enjoyed in the past is now in turbulence.

Secrets have been revealed. The unthinkable is exposed by proof and evidence. Time we once knew is no more.

Three ... Two ... One ... Blastoff

Prepare for the awakening of outer space exploration as our planet Earth Space Federation Council joins the extraterrestrial Galactic Federation.

Space 2
Extraterrestrial

Extraterrestrial implies originating, existing, or occurring outside of the atmosphere of planet Earth. Any beings not from earth or humans traveling to other worlds are classified as Extraterrestrial.

Want to know the secrets of the universe? Want to live among the stars and see what no one on Earth has seen before?

We can begin our journey to the Moon and stars right here. We are about to explore the universe together taking the first step.

There are those who dare and those who dare not. The call for a planetary defense initiative begins and ends with an SIA Space Intelligence Agency. So let our story begin ...

There are an estimated 60 billion planets in our galaxy, 200 billion observable galaxies and another foreseeable 500 galaxies in the universe. There are a projected 17 billion Earth size planets in our galaxy. There are about 100 thousand million stars just in our Milky Way. It is estimated there are 40 sextillion (40,000,000,000,000,000,000,000) stars in the universe.

Estimating expansion and continuing expansion of the universe is thought to be 13.82 billion years old. No one knows for sure. At present, we can only see into the universe 1-to-2 trillion years. Anyway, you get the picture. It's BIG out there.

Outer Space

Infinity without defined boundaries of existence beyond the atmosphere of Earth where Earthlings occupy its terrain is composed of solar systems of celestial bodies of stars and planets.

The purpose that drives us is our warrior nature to conquer the unknown and colonize unoccupied planets. Whether in a partnership or solely operating in the exploration phase or development stage on other planets, the lack of a central governing body for Earth places all of us in jeopardy.

For things to work there will come a time for having only one established Earth Space Federation Council with sole jurisdiction and authority over all of Earth's extraterrestrial activities. This is leading up to eventually having one federation military force and one federation intelligence agency in a united defense of planet Earth and space exploration. The idea is not the New World Order having one country and a few politicians controlling the world and its people.

Multiple representations on planet Earth in the extraterrestrial will not work either. I intend to lay the groundwork for the first extraterrestrial spy agency.

Outer space exists in many different layers. Space is a vacuum within a vacuum of low-density hydrogen and helium plasma. Space is made up of subatomic to microscopic particles. They vary in properties of volume, density or mass moving in a crowd or as celestial bodies in motion emanating electromagnetic radiation, magnetic fields, neutrinos, dust, and cosmic rays.

The baseline temperature of outer space is 2.7 kelvins (-270.45°C, -458.81°F). Intergalactic space makes up most of the volume of the universe. Even galaxies and star systems consist almost entirely of empty space. Studies show 90% of the mass in most galaxies is in an unknown form referred to as dark matter.

It appears we are just getting started into research and discovery in the sciences of outer space There is no reason for outer space not to be part of our tomorrow.

Definitions and Terminology Related to Outer Space
The following are key words and terms of importance to better understand the universe as it is currently defined.

From Dimensions to Outer Space
A boundless three-dimensional extent of distance and time beyond the atmosphere of planet Earth.

Geospace
The region of outer space closest to planet Earth.

Cislunar Space
The region of outer space outside of Geospace extending just beyond the orbit of the Moon.

Solar System
The gravitationally bound system of the Sun and the planets around it. Planet Earth is in a younger solar system of 4.425 billion years old.

Deep Space
The region of outer space beyond the Moon throughout the universe.

Interplanetary Space
The region of space within our Solar System.

Interstellar Space
The physical space within a galaxy beyond the atmosphere of a star. This touches the Stellar Winds which gradually erode the outer atmosphere of stars.

Intergalactic Space
The physical space between galaxies.

Spacetime
Time makes up the boundless fourth-dimension continuum.

Dimensions
The range or degree over which something extends in proportion or size is fundamental to an understanding of the physical universe ...
First dimension ... width (line)
Second dimension ... height (plane)
Third dimension ... depth (space)
Fourth dimension ... time (Einstein theory of relativity)
Fifth dimension ... similarities and differences (variances)
Sixth dimension ... past, present, future (static universe)
Seventh dimension ... gravity and light (invisible universe)
Eighth dimension ... past and future infinities (existence)
Ninth dimension ... all possible pasts (thoughts)
Tenth dimension ... all possible futures (comings)
Eleventh dimension ... infinities come together (reality).

Dimensions are considered the theory of everything. A multiverse exists in gravity, light, and temperature unless broken into dimensions of time and space. We can see the first three dimensions. The others are invisible to us.

While the idea of a multiverse is revered, it seems to remain amusement in terms of science fiction and daydreaming. Though convincingly sound mathematically, it brings together a beautiful mind in science and advanced combinations of numbers.

At this time there are two basic theories. One is the original Bosonic string theory of the 1960s. The other is the Supersymmetry string theory of the 1980s. The Bosonic string theory is spectrum-based on bosons or a particle such as a photon whose number is zero or an integral number. The Supersymmetry string theory is spectrum-based on fermions or composite particles of electrons, protons, neutrons.

The point being string theory is the more popular basic thought suggesting there are only 11 dimensions existing with which we can attempt to understand our world and the universe. Space-minded scientists then believe it is possible using telescopes to observe light from the early universe which allows us to peer back through time. Far out stuff.

The Cosmos
The Cosmos is the self-inclusive system of things in the universe.

Galaxy
A galaxy is a large group of stars.

In our galaxy there are around 400 billion stars and 400 billion planets. There are around 200 billion galaxies in the universe.

Milky Way
The Milky Way Galaxy contains the solar system of planet Earth. Scientists have calculated there are 6-billion Earth-like planets in the Milky Way Galaxy.

Stars
Stars shine by giving off their own light and every star has at least one planet.

Scientists have calculated there are an estimated 400-billion stars in the Milky Way Galaxy.

Exoplanets
Exoplanets are the planets outside of the solar system of Earth. Scientists have confirmed the existence of 4,164 exoplanets. There are thousands of other candidate exoplanet detections that require further observation.

Protoplanets
Small size planets are up to 1,200 miles in diameter.

Planets
Planets reflect light. The planet Proxima Centauri b orbiting the star Proxima Centauri is believed in the closest habitable zone.

Scientists using space telescopes have identified 3,750 planets. Planets will eventually house power stations for refueling and supplies. Some will become outposts and others new civilizations.

The environment of a planet determines what life would look like on each planet. The measure of a planet's habitability is an extrapolation of the conditions on planet Earth. These same characteristics with the potential to sustain hospitable life on another planet would ideally have elements of water, oxygen, and soil with cellular nutrients.

What the Observable Universe Looks Like

Very little is known about the universe. However, we do know that it is expanding continuously.

Dark matter and dark energy are the cosmic fuel powering the expansion of the universe. The dark surrounding area is made up of 26 percent dark matter believed to be galaxy clusters. The cosmic fuel accelerating the expansion of the universe is made up of 70 percent dark energy. The other four percent is of an unknown quantity.

The ingredients of this catalyst remain a mystery. In other words, 96 percent of the universe lacks understanding. It remains a question still as to the cause of the acceleration in the expansion of the universe.

On 13 July 2019 scientists launched an astrophysics venture to map the universe. The Spektr-RG (Spektrum-Röntgen-Gamma) astronomy orbital telescope was sent 1.5-million km from Earth to map the radiation as seen through x-rays of the cosmos.

The German-developed eRosita (Extended Roentgen Survey with an Imaging Telescope Array) instrument system was mounted on the Russian-built ART-XC (Astronomical Roentgen Telescope — X-ray Concentrator) science hardware. I have to wonder why the expanse seems to stay one element out of the reach of Earthlings ...?

Existing Civilizations in the Milky Way Galaxy
Astronomer Astrophysicist Frank Drake was the first to map our Milky Way Galaxy.

The overlap between the lifetime of our civilization and the number of extraterrestrial civilizations in our Milky Way Galaxy will determine the best means of transportation and manipulation of time travel. We are talking about very low probabilities; we will find nothing.

The Drake equation is used to estimate the number of active communicative extraterrestrial civilizations in our Milky Way Galaxy.

$N = R_* f_p n_e f_l f_i f_c L$

Where ...
N = the number of civilizations in the Milky Way Galaxy whose electromagnetic emissions are detectable with which communication might be possible
and
R_* = the average rate of star formation in our Milky Way Galaxy suitable for the development of intelligent life
f_p = the fraction of those stars that have planets with planetary systems
n_e = the average number of planets per solar system with an environment suitable for potentially supporting life
f_l = the fraction of suitable planets on which life actually appears or could support the actual development of life at some point
f_i = the fraction of life bearing planets on which extraterrestrial intelligent life emerges and actually go on to develop intelligent extraterrestrial life civilizations
f_c = the fraction of extraterrestrial civilizations that develop a technology capable of releasing detectable signs of their existence into outer space
L = the length of time for which such extraterrestrial civilizations release detectable signals into outer space

Conclusion 15 June 2020 published in the Astrophysical Journal ...

36 extraterrestrial civilizations exist in our milky way right now.

Let's consider what we are being told ...

What you are about to encounter is real. Welcome to the new reality of outer space contact and travel.

From civilian to commercial and military to SIA Space Intelligence Agency the era of Earth security and defense has begun with this book. There are a few intelligence agencies focused on the extraterrestrial. However, no extraterrestrial intelligence agency exists at this time for any government or military in the world. Taking the lead in any new endeavor always goes against the forces of doubt and criticism.

The following brief composite of isolated people and events is only a minor introduction of alien extraterrestrial encounters to what governments and departments of defense across the world have admitted, both verified and documented in precise detail.

What makes something important? It changes your life.

Is it more important to know why something happens? Probably not unless it affects your life. Why is this important? To better understand the new realities it means to be informed.

As for extraterrestrials, we ask who are they? When did they first come? Why did they come? Who stayed? What did they leave behind? Where are they? Are they friend or foe?

While interesting, what you are about to read is important in laying the foundation of necessity for a space intelligence agency.

It is important what you think about what you have read, what you accept and what you reject. Our future lies with us.

The course of what we have been given contains an obligation to provide security and defense of our planet. The only way to preserve what we think we know is to understand what we don't.

For those who hesitate all may be lost.

Sensitive Classified Intelligence (SCI)
SCI Sensitive Classified Intelligence, now Declassified, presents an interesting scenario of past revelations. The standard period most things remain classified is 25-years.

It is not fear, suppressing evidence of encounters with the unknown and little understood, but ignorance.

Forget about what you have been told to think about in the past. Explanations of alien extraterrestrials encounters have been from the extreme to sketchy at most. Think about what might be possible.

Outer Space Aliens
Extraterrestrials are non-human intelligent beings.

Governments and militaries confirm former visitations of ancient extraterrestrials and living extraterrestrial aliens on Earth.

Outdated Intelligence
Knowledge of existing extraterrestrial aliens makes every intelligence service and agency outdated.

Extraterrestrials ... What's Real and What's Not
Keep an open mind ... the next few pages may blur what you have thought was not true with what you might have suspected.

It is human nature and behavior to reject all this as nonsense. People in bureaucratic environments want to do what they have always done and guard what they have always defended in the status quo of the past. However, change is upon us.

As for you, the only way you are going to believe is if you have a first-hand encounter with the future. Human conditioning says so.

Do alien extraterrestrials come in peace or do they come in war?

Maybe both. It makes no sense to think we do not have ET alien enemies from outer space. To best answer this question let's journey from the present of what we do know into the future.

Extraterrestrial Alien Forms of Communications
Non-verbal, telepathy, and speech communication forms are used to message with each other and used by ET life forms and human beings.

Mental telepathy, alien languages, and every language spoken on planet Earth are used to communicate. ET aliens have a telepathy open and closed loop with each other for mass communication and privacy.

Language
Speaking without words is acceptable.

Diversity of life is often confused for advancement toward singularity. Mental images, thoughts, and feelings are the basic form of alien conversations with nonaliens. Classified alien dialogues continue.

Reportedly at this time a common single interstellar language is under development with extraterrestrial alien participation.

Future Events
It is undetermined if extraterrestrial alien forecasts of future events for planet Earth are viable forecasts or manipulated history.

Spacetime
Time can bend like light. However, time does not always travel at the same speed as light. This being ... time in space already exists when light comes into existence.

Time Warp
Time warp is a discontinuity, suspension, or manipulation of time in spatial dilation or contraction.

Warp Drive
Warp drive is the ability to travel faster than light-speed.

Alcubierre drive is a product of Einstein's field equations in general relativity by which a spacecraft can travel faster than light with expanding and contracting spacetime in the vacuum of outer space. We shall see because ...

We Are Looking

What we don't know can hurt us and when we do know we may not be able to immediately do anything about it.

Space 3
Extraterrestrial Alien Life Forms on Earth

Earth Home to Extraterrestrial Alien Life Forms
Verified extraterrestrial life forms on planet Earth are acknowledged by governments and militaries.

Alien body forms and human forms of extraterrestrial life forms exist residing within a mixture of sea, rock bases, and among the human population.

Chile has more UFO Unidentified Flying Object sightings than any country in the world.

Chile has confirmed with data and evidence UFO visitations of extraterrestrial alien spacecraft from other planets.

Chile
Chile is one of the largest uninhabited areas on planet Earth.

The government and military of Chile confirms visitations of extraterrestrial alien aliases from other planets.

In their investigations, they have discovered it is possible to create frozen-time frames of life.

This was confirmed with the abduction of a Chilean military soldier who was lost for a matter of Earth minutes and returned with a 5-day old beard.

Chile, having enormous amounts of raw materials and rare Earth elements, is ideal for extraterrestrial bases and flight paths of UFOs coming to planet Earth.

Extraterrestrial Alien Spacecraft
UFOs, flying saucers, various spacecraft, Moon bases, and one-year light travel claims exist.

Well documented and reversed engineered are all alien craft and technologies for protection and prevention of intergalactic war and advancement of space travel.

Mapping Outer Space
The Earth was flat and the cosmos was round and now it is neither.

3-D cosmic mapping the infinite requires state-of-the-art science. We know neither the age nor the time of the cosmos to be explicit.

Galaxies make the stars. Stars make the planets. Planets and Moons make life. The universe is glued together by centrifugal forces.

God Particle
The God particle to accelerate life on another planet will provide all the elements on Earth and support life as it exists.

To colonize another planet it is necessary to make that planet habitable for Earthlings. This biomarker bomb can explode on another planet and within a defined time certain be ready for occupation.

Time Travel
At the speed of light time slows down.

At MIT an experiment was conducted and a worldwide collection of notable research think tanks were able to replicate and confirm the movement of matter at three times the speed of light.

The acceleration of light speed in Earth's atmosphere within a laboratory setting at MIT has evolved a new equation of particle matter transfer through space. Element 115 gravity warp drive was a byproduct. Its byproduct application is inter-dimensional travel.

As light bends research is expanding this breakthrough for the exploration of outer space through stargates, black holes, and wormholes. This has opened the missing element for time travel which may allow speeding through the galaxies to specified designations in a matter of minutes, hours and days.

Outer space is not a perfect vacuum as it contains particle matter. Current thinking is time travel changes time, and if so, there exists a time discrepancy.

Genetic Engineering of Intergalactic Beings
Genetic engineered advancement implementation continues to evolve the human condition for outer space travel.

Bio-genetic babies are given birth-bodies for adaptation to survive and occupy a common existence on other planets. This makes for an invisible spy that can blend in and infiltrate any species or hierarchy on other planets to conduct the spycraft of espionage.

Look around and take notice of the size in height of humans today. It might make you think we are living in the land of giants.

Scientists have identified an alien chromosome believed implanted for selective breeding of humans with certain traits. To date it has given birth to 5'10" – 6'5" women and 6'5"-7'5" men. Medical reports confirm the genetic evolution of the next generation may produce a 10'-12' species of superior beings.

Some researchers believe the immortal jellyfish gene is being intentionally withheld from embryonic implant known to give eternal life. It may not guarantee you will live forever however it may very well extend the lifespan we have come to accept.

Earth will continue to be a breeding ground for new species and extraterrestrial alien genetic experimentation of DNA and genomic blending. This is important in the advancement of science and the evolution of the human race.

More importantly experimentation may very well allow us to create a species of Earthlings capable of outer space travel without the need for current life-support systems.

One way or the other we will continue to progress our adaptation for space travel and exploration. It will not be an extension of Earth but rather a need breed of explorer for the cosmos.

Future settlements on other planets will be colonized by a species of the best Earthlings we have to offer the future. I expect at some point they will separate from their motherly attachment to Earth.

Preparation of Earth Beings for Outer Space Travel
Hybrid humans in the future will have gene editing by molecular engineering techniques for space travel.

Imagine a species of Earth living on another planet that has no oxygen or a planet with hostile bacteria or viruses. There may even exist planets with an unknown quantity capable of altering genetics of beings that could replicate us into threats.

Where scientists lead the way, it is technology that gives our capacity to understand. The forbidden will always be there and we do not want to bring it home.

For us to travel the universe with no space suits for breathing makes us better prepared to enter alien environments and then exit them. No gravity requirement for walking. No food necessity for energy needed for extended periods. No water demand for life maintenance. Then again what is sci-fi today may become tomorrow's reality as ...

Suspend time for evaluating a situation. Medical scanner for self-medical diagnosis and cure. Weapons to stun or defend an enemy threat or attacker. Having the ability to be self-sufficient if alone or abandoned. When we think about it, it all comes down to survival.

Preparation precedes participation.

Extraterrestrial Alien Encounters
All alien encounters are identified and classified or Declassified.

Governments acknowledge the assimilation of many types of extraterrestrial aliens married to humans, spreading the alien gene, and working among the population of the world. This admission supports the existence of all types of phenomena. It does not matter what we think or believe as our role of existence will probably not change.

Only a small cadre of individuals are aware of the situation. Now you are too.

Close Encounters of the Third Kind

The first recorded and fully documented modern extraterrestrial alien encounter was presented to the public in the movie *Close Encounters of the Third Kind*.

There is an interesting story of movie director Stephen Spielberg afloat. Authorization with the encouragement of the US government was for a movie to be made about an actual UFO contact event.

Myth or lore, the movie was produced. For many of us, our life experiences come from movies and television. Because what we see and hear oftentimes gets mixed up with our conscience and subconscious. Our reality can easily become confused with fantasy.

Governments know. Now whether they decide to share what they know is altogether another issue.

All governments have a primary interest in their survival and continued existence. They will do whatever is necessary to prepare their population to receive or accept their opinion or change.

Foresight was to begin the conditioning process of the public on extraterrestrial alien life and preparation for open contact beginning with this movie. This was a pretty big gesture on the part of the military and those holding higher government positions.

In ufology, a close encounter is an event in which an Earthling witnesses a UFO occurrence while seeing an ET extraterrestrial alien not from planet Earth and/or having an intelligent communication.

Dr. J. Allen Hynek was the personal consultant to Steven Spielberg for the popular 1977 movie *Close Encounters of the Third Kind*. The name of this classic extraterrestrial movie about an actual ET and UFO event was named after one of the levels of Hynek's scale.

We find a blending of fact and fiction about ETs and UFOs presented to the world for the first time. This revealing true story of an extraterrestrial phenomenon finds Hynek making a cameo appearance at the end of the film stepping forward in beard with pipe in mouth.

Hynek System of ET and UFO Sighting Reports Classifications

It was September 1947 when the US Air Force sought the help of Dr. J. Allen Hynek to develop a scale of classifications to better catalog ET and UFO extraterrestrial phenomenon reports known as Project Sign.

During the first year of Project Sign, Hynek reviewed 237 ET and UFO cases. His classification system attributed 32 percent of incidents to astronomical phenomena, 35 percent of incidents to balloons, rockets, flares or birds, 13 percent of incidents were without sufficient evidence to even yield an explanation, and the remaining 20 percent of UFO incidents provided some evidence and couldn't be explained.

The Hynek paranormal program provided the needed structured system when investigating ET extraterrestrial alien and UFO sightings in the past. As for the future, the Hynek and Valle systems provide the two current systems widely used by ufologists, governments, and militaries.

Dr. J. Allen Hynek was the first astronomer who first classified ETs and UFOs. The original work of Hynek continues to help easily explain which cases of astronomical phenomena are planets, stars, or meteors.

In 1949 Project Sign was succeeded by Project Grudge. Project Sign offered a pretense of scientific objectivity. From the start Project Grudge as its name suggested was dismissive and took the premise ET extraterrestrial aliens and UFOs cannot exist. The intent was to deny.

Project Grudge recommended the subject of ET extraterrestrial aliens and UFOs wasn't worth further study. It might have ended there.

As ET alien contacts and UFO sightings continued, including government and military reports of such, Project Blue Book was born. In 1952 Hynek joined project Blue Book until its demise in 1969.

The denials leading to mistreatment of witnesses as lying, insane, hallucinating collectively by normal people gave reluctance for others to come forward. The result was a significant loss of data to researchers and scientists.

The Hynek System

Dr. J. Allen Hynek was the first pioneer of classified ET and UFO sightings into categories of distant classifications and kinds of close encounters.

Distant Classifications

NL Nocturnal Light ... A simple visual sighting of an unidentified flying light seen at night.

ND Nocturnal Disc ... A simple visual sighting of an unidentified flying extended or structured light source seen at night.

DD Daylight Disc ... A simple visual sighting of a UFO with distinct oval or discoidal shape seen during the day.

Radar-Visual Reports ... UFOs observed visually whilst being seen on radar and/or UFO simultaneously seen in person. Radar propagation can be occasionally discredited due to atmospheric propagation anomalies. Hynek determined these make up 1 to 2 percent of reports.

Close Encounters

Close Encounters of the First Kind (CE1, CEI) ... A CE1 is an observation of a UFO within 500 feet or 165 yards or 150 meters. Close enough to show an appreciable angular extension and considerable detail greatly reducing or eliminating the possibility of misidentifying conventional aircraft or other known phenomena.

Close Encounters of the Second Kind (CE2, CEII) ... A UFO which leaves some form of physical evidence. It can be broken grass or radiation or the finding of material of unknown makeup. Physical effects remain.

This can be interference in the functioning of a vehicle or electronic device, animals reacting, a physiological effect such as paralysis or heat and discomfort in the witness, or some physical trace like impressions in the ground, scorched or otherwise affected vegetation, or a chemical trace, or radiation burns.

Close Encounters of the Third Kind (CE3, CEIII) ... A visual sighting of an ET occupant or alien entity associated with a UFO. These include humanoids, robots, and humans who seem to be occupants or pilots of a UFO. An analysis by Hynek of 650 reports found only 1% to be CE3.

An entity is observed only inside the UFO.
An entity is observed inside and outside the UFO.
An entity is observed near to a UFO but not going in or out.
An entity is observed and no UFO is seen by the observer but UFO activity has been reported in the area at about the same time.
An entity is observed but no UFO is seen and no UFO activity is reported in the area at that time.
No entity or UFO is observed but the subject-witness experiences some kind of intelligent communication.

Close Encounters of the Fourth Kind (CE4, CEIV) ... An abduction of a human by an ET alien being or species often including experiments done by the UFO occupants. These Earthlings taken experienced a transformation of their sense of reality. This also includes non-abduction events where absurd, hallucinatory, dreamlike happenings occurred with UFO encounters.

Close Encounters of the Fifth Kind (CE5, CEV) ... A direct cooperative contact and communication with a friendly ET extraterrestrial alien or species. This occurs through conscious, voluntary, and proactive human-initiated experience with an extraterrestrial intelligence.

Close Encounters of the Sixth Kind (CE6, CEVI) ... Human or animal deaths associated with a UFO sighting. This could be considered a more extreme example of the second kind.

Close Encounters of the Seventh Kind (CE7, CEVII) ... A Human/ET hybridization. The creation of a human-ET hybrid child either by reproduction or by artificial scientific methods.

Astronomer, Ufologist, Astrophysicist
Dr J. Allen Hynek the Father of Ufology

Vallee System of ET and UFO Sighting Reports Classifications
The Dr. Jacques Vallee System was invented as a more in-depth system for reporting and classifying phenomena and paranormal experiences of ETs and UFOs visiting Earth.

Vallee saw objects from outer space not man-made while serving on the French Space Committee. He also witnessed proof and evidence of such destroyed by the French government.

In 1963 Dr. Jacques Vallee co-developed the first computerized map of Mars for NASA.

The breakdown in investigations of ETs and UFOs visiting Earth classified phenomenon into AN ratings.

AN RATING ... Classifies any anomalous behavior.

AN1 Anomalies which have no lasting physical effects. i.e. amorphous lights, unexplained explosions.

AN2 Anomalies which do have lasting physical effects. i.e. poltergeists, materialized objects, areas of flattened grass, corn circles.

AN3 Anomalies with associated entities. i.e. ghosts, yetis, spirits, elves, and other mythical or legendary entities.

AN4 Witness interaction with the AN3 entities. i.e. near-death experiences, religious miracles and visions, OBEs out-of-body experiences.

AN5 Anomalous reports of injuries and deaths. i.e. SHC Spontaneous Human Combustion of unexplained wounds as well as permanent healing that results from a paranormal experience.

MA RATING Describes the behavior of a UFO. It is analogous to the Nocturnal Light, Daylight Disk, and Radar Visual Hynek classifications.

MA1 A UFO has been observed which travels in a discontinuous trajectory. i.e. vertical drops, maneuvers, or loops.

MA2 MA1 plus any physical effects caused by the UFO.

MA3 MA1 plus any entities observed on board. i.e. the airship cases of the late nineteenth century.

MA4 Maneuvers accompanied by a sense of reality transformation for the observer.

MA5 A maneuver that results in a permanent injury or death of the witness.

FB RATING Fly-by rating.

FB1 A simple sighting of a UFO traveling in a straight line across the sky.

FB2 FB1 accompanied by physical evidence.

FB3 A fly-by where entities are observed on board, which is rare.

FB4 A fly-by in where the witness experienced a transformation of reality into the object or its occupants.

FB5 A fly-by which the witness would suffer permanent injuries or even death.

CE RATING Close Encounter ratings.

CE1 UFO comes within 500 feet of the witness, but no aftereffects are suffered by the witness or the surrounding area.

CE2 A CE1 that leaves landing traces or injuries to the witness.

CE3 Entities have been observed on the UFO.

CE4 The witness has been abducted.

CE5 CE4 which results in permanent psychological injuries or death.

Computer Scientist, Ufologist, Astronomer, and Astrophysicist Dr. Jacques Vallee the Grandfather of Ufology in 1955 and again in 1971 saw a UFO and witnessed the tapes of evidence destroyed.

Ufology
The term Ufology is related to the study of the term UFO.

The term UFO Unidentified Flying Object was coined in 1953 by US Air Force officer Edward Ruppelt the director of Project Grudge and Project Blue Book. In accordance with US Air Force Regulation 200-2, the term UFO was in the beginning restricted to the fraction of cases which remained unidentified after investigation.

The term Ufology is the study of unidentified flying objects. From pseudoscience nonsense of skeptics to the established scientific community of phenomena each has its unique contributions.

Ufologist or UFO Investigator examines the proof and evidence of ETs and UFOs sighting and experiences. Where science is the study of the physical universe and ufology is the study of unexplained phenomena makes both science.

Ufologists and scientists examine cases of ETs and UFOs reported, interviews, data collections of information, pictures, and videos. Research indicates a third of Americans believe in extraterrestrial ETs and UFOs. Whichever the case, ET and UFO sightings continue as ...

They Live Among Us

We should be thankful for genetic advancement, intelligence evolution, new knowledge and protection of planet Earth.

Space 4
Extraterrestrial Law and Jurisdiction
International Business Law for the 21st Century Executive (Tom Grooms J.D. 1991)

Outer Space Common Ground
Earth has no legal jurisdiction beyond the atmosphere of planet Earth. National sovereignty does not extend beyond Earth. Outer space doesn't belong to any one country, group of countries, or any organization usurping authority. Outer space is common ground.

There is no existing governing body of planet Earth having the authority to codify laws, exercise police powers, and form a tribunal to hear disputes on matters related to outer space. In 1958 UNOOSA the United Nations Office for Outer Space Affairs was established to promote peaceful cohabitation among member states in outer space.

The United Nations has no legal jurisdictional standing to hear or try cases related to the extraterrestrial, establish enforceable legal and regulatory frameworks to govern outer space activities, or authority to grant or deny access to the use and exploration of the extraterrestrial beyond the atmosphere of Earth. No existing treaty between any of the countries has legal jurisdiction beyond the atmosphere of planet Earth.

Space-participating nations on planet Earth coexist by agreements of treaty between each other, hopefully peacefully, in outer space.

Outer space is open territory until occupied. Planets are for colonization. Therefore outer space remains an open frontier for exploitative and development activity.

Interplanetary Space Law
As the inhabitants of Earth extend for the stars, they find others who have gone before. While others have already been, there are those waiting to travel beyond the unknown. Extraterrestrial occupants on other planets will define interplanetary jurisprudence space law.

Individual rights and legal defenses on other planets will be determined by interplanetary law and ancient codes of existing federations. Federation jurisdictions have extraterrestrial rule.

Outer space exists among a confederation of planets.

Extraterrestrial Jurisprudence
A confederation of planets has a willing system and body of extraterrestrial jurisprudence based on contract law and criminal law.

The situs of outer space law regulates the relationships between differing extraterrestrial species of beings in the universe. No one planet has the jurisdiction to codify laws of outer space.

However, each planet has the situs of jurisprudence to codify laws and regulations specific to that planet. For any extraterrestrial planet to implement internal law it must take into account the much larger political and social context. Legal reasoning, legal systems and legal institutions must conform accordingly for the role of law in any society.

Extraterrestrial Contract Law
Extraterrestrial contract law is doing all of what you promised to do by keeping your word both oral and written.

This legal principle applies to the sale of goods, property rights and the conduct of trade emphasizing the special features of contract performance due to its extraterrestrial character. A body of extraterrestrial tort law evolves on case precedents of civil rulings.

Legal issues of consideration regarding extraterrestrial carriage of goods, materials and species between planets will require special rules governing transports. Physical risk of damaged, delayed or missing goods in transit customs and practices figure prominently in rulings. Payment in extraterrestrial sales require new mechanisms of finance and credit between buyers and sellers and third parties.

Extraterrestrial Criminal Law
Extraterrestrial criminal law is not injuring other extraterrestrials or beings, or unlawfully taking, damaging or destroying their property.

In other words destructive violations of extraterrestrials and their property is a criminal act. A de facto jurisprudence body charged with hearing cases in extraterrestrial criminal law will require wide latitude in determining guilt or innocence. A body of extraterrestrial criminal law evolves on case precedents of procedures and rulings.

Space 5
Revealing Sensitive Declassified ET and UFO Documents

Sensitive Declassified Intelligence ET Encounters and UFO Sightings

There are many notable extraterrestrial alien encounters that provide a convincing argument for a Space Intelligence Agency.

Anytime extraterrestrial aliens or UFOs are mentioned conspiracy theories, labels, and accusations fly. The way the topic has been treated makes it sensitive.

Character assassination is easily achieved by labeling someone having a conspiracy theory. This means the person has no evidence or proof for what they believe to be true. All denied truth is treated as a conspiracy theory. This happens when someone accuses a government, institution, or person of denying what it knows to be true. In turn, this also happens when a person, organization, or institution has their credibility destroyed regarding a claim known to be true that cannot be proven.

It seems extraterrestrial aliens and UFOs fall into this category of conspiracies, accusations, and denials in both directions from both accusers and vindicators. However, the matter is essential to the subject of this book and will be discussed in some length.

It was interesting to review reported and recorded extraterrestrial encounters admitted by governments and their militaries throughout the world. UFO sightings appear concentrated on monitoring military bases and nuclear test sights. Throughout the world human and non-human intelligence collections of phenomena are conducted by UFOs.

The following list of extraterrestrial and UFO encounters are the best-substantiated, documented and publicly acknowledged events. Research of extraterrestrial aliens and UFO occurrences are included to inform and establish the foundation for a Space Intelligence Agency.

What does the past leave behind when the future calls ... ?

Extraterrestrial Alien Encounters and UFO Sightings & Incidents
The following sensitive Declassified intelligence is not all-inclusive or final by any means. These are derived from credible secondary sources. That is not all. Draw your own conclusions.

5000 BC Egypt Era of Tezeppi referred to as the beginning of our time cited extraterrestrial alien visits and UFOs.

4500 BC China extraterrestrial UFO flying Dragon from the star Chi.

5 BC to 2 BC Maha Bharata, India radiological evidence of nuclear weapons was described in historical records of Mahabharata Sanskrit.

1492 October 11 sightings of UFOs out of the sea in the Bermuda Triangle believed by some to be the past colony of UFO headquarters.

1561 April 14 Nuremberg, Germany the entire town of thousands at dawn witnessed an ET celestial war. A battle of major proportion reportedly filled the sky. It was recorded on parchment that still exists. This major battle was fought by different shaped UFOs overhead. This is the first record of mass sightings.

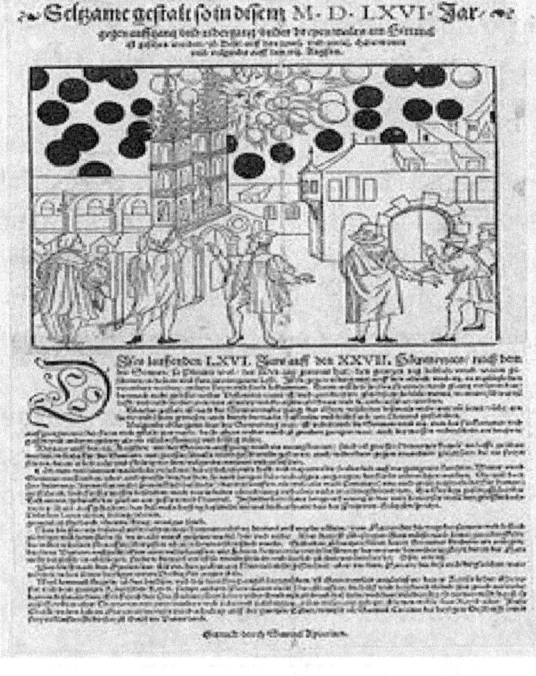

In 1566 the Basel Pamphlet of Switzerland published accounts of the two days at sunrise and sunset on the 27-28 of July and on August 7 of ET-UFO wars which occurred in the skies like on April 14. The wood pamphlet that still exists revealed every type and variety of military-shaped UFO spacecraft in likeness as reported today.

United States and UFOs

1639 March United States first reported UFO sighting by 3 men in a boat on Muddy River in Massachusetts.

1728 Philadelphia, Pennsylvania Benjamin Franklin believed in extraterrestrial life on other planets and wrote of his scientific view.

1776 United States first UFO sighting cited in Winthrop's Journal of New England in 1790.

1777 Winter Valley Forge, Pennsylvania President George Washington UFO encounter.

1800 April 5 Natchez, Baton Rouge, Louisiana Vice President Thomas Jefferson UFO encounter.

1865 April 14 President Abe Lincoln, a few hours before his assassination, admitted to General Ulysses Grant about his encounter with a UFO.

1870 Mount Washington cylinder-shaped UFO photographed.

1883 Marfa Texas Lights about nine miles east on highway 67 is a mysterious phenomenon that the world's leading scientists have not been able to solve.

Your author saw the lights first-hand while sitting on a fence post at 1:00 am when visiting the area in southwest Texas. I wondered why here at this location would a phenomena event like this take place.

It appeared to me like flashlights going off in rhythmical succession and not in sequence or any pattern. I thought maybe this was some kind of gaseous light energy like the northern lights but the form of lights flashing was circular in shape.

Guess it is good there are still things in our world that cannot be explained giving a sense of humanness frailty.

1896 June 10 American settlement moving west reported waves of UFO sightings. Remember it was 1903 when the Wright Brothers took their first flight.

1897 April 17 Aurora, Texas 30-miles northwest of Fort Worth in *The News* reported a cigar-shaped space craft had collided with a windmill. The crash scattered extraterrestrial craft metal debris over many acres.

The extraterrestrial alien onboard was not of this world. The papers in unknown hieroglyphics found on the body of the extraterrestrial pilot could not be deciphered.

1907 August 1 Sagamore Hill, Oster Bay Long Island, New York 9:00p-11:00p President Theodore Roosevelt at estate home witnessed a UFO.

1922 Mohenjo-Daro, Pakistan was discovered. There was evidence of radiation from atomic warfare and an abandoned city of 2600-1900 BC.

1923 Rome, Italy while researching at Vatican Secret Archives Russian scientist Genrikh Ludvig confirmed by photographs and manuscripts evidence of extraterrestrial aliens visiting Earth, spaceships, and UFOs.

1942 USAF Research Base at Groom Lake established. At a much later date an S-4 Facility hides extraterrestrial alien technology of 9 UFOs.

1945 July 16 Alamogordo, New Mexico Trinity Nuclear Test Site disturbs ETs and brought green fire-ball drones from a UFO mother ship in outer space.

Later on, the same events occurred at Los Alamos National Laboratory near Santa Fe, New Mexico - Sandia National Laboratories at Albuquerque, New Mexico - Savannah River National Laboratory near Jackson, South Carolina and Oak Ridge Laboratory in Tennessee.

1947 June 21 Maury Island Marine Park in Washington state UFO incident of 6 UFOs above a boat and first encounter with Men in Black.

1947 June 24 state of Washington Hanford Nuclear Facility the first US plutonium production facility experienced 9 UFOs reconnaissance.

1947 July 2 Roswell New Mexico UFO Crash
Rancher W.W. Mac Brazel discovered a large area of UFO bright material wreckage. The material was extraterrestrial and its properties were not of planet Earth.

The Roswell Daily Record reported the Army found a flying saucer. The UFO crash was attributed to an electrical lightning storm.

Roswell Army Airfield Major Jesse Marcel was sent to investigate. The War Department claimed it was a balloon from the Top-Secret Project Mogul which monitored Russian nuclear testing.

Pilot Oliver Pappy Henderson saw 4 extraterrestrial alien bodies at the Roswell base. An autopsy was performed and recorded.

1947 July 5 reportedly 4 UFO four-foot extraterrestrial alien cadaver bodies were flown to Fairborn Ohio Wright-Patterson Air Force Base. One surviving extraterrestrial alien was flown to Lowry Field and later taken to Wright-Patterson Air Force Base.

Roswell Daily Record 8 July 1947 .... Associated Press 9 July 1947

Aftermath of Roswell
Lt. Col. Richard French reported there was a second UFO crash near Roswell within a few days after the first one. The second scout UFO was looking for survivors. CIA Chase Brandon confirmed CIA materials and pictures evidenced the UFOs were not from this planet.

4602D Air Intelligence Service Squadron
In 1952 March 1 the 4602D Air Intelligence Service Squadron created the first covert military collection agency of the US Air Force. It was designated by the Air Defense Command to by-pass Project Blue Book in the official investigations of extraterrestrial aliens and UFOs.

All reports of extraterrestrial aliens and UFOs were to go through the 4602D before any transmission to Project Blue Book. The 4602D dealt with more sensitive cases of national security concerns.

The 4602D was designated Project Operation Ivy. Their specialty was monitoring nuclear weapons tests. Their squadron was headquartered at Ent Air Force Base in Colorado Springs, Colorado and later moved to Fort Belvoir, Virginia.

Each 4602D intelligence team consisted of 3 men. One linguist, one technological man, and one special operations man. All 3 were airborne qualified and cross-trained.

The 4602D had three missions ...

UFO Program for investigation of reliably reported extraterrestrial aliens and unidentified flying objects.

Project Moondust to locate, receive, recover, and deliver descended foreign vehicles from space, especially of non-US space technology, extraterrestrial alien spacecraft and objects of unknown origin used in examination of phenomena.

Project Blue Fly was a special unit with a focus to facilitate the identification and collection of phenomena for expeditious delivery to the Foreign Technological Division of Moondust items of great technical intelligence interest.

The team mission critical was to arrive first at any site within 6-hours anywhere in the world.

Project Blue Book
In 1947 Project Sign began the first known official government investigations of UFOs. In 1949 this became known as Project Grudge.

In 1951 May 21 the designated headquarters for Project Blue Book was the Air Technical Intelligence Center at Wright-Patterson Air Force Base in Dayton Ohio.

Project Blue Book had two primary missions ...
1. To determine if UFOs were a threat to national security.
2. To scientifically analyze UFO-related evidence and data.

From 1948-1969 there were 12,618 UFO investigations and reports acknowledged. In 1947 there were 800+ UFO investigations and reports written. J Allen Hynek

In December 1969 Project Blue Book was terminated.

In 1970 the Department of Defense assumed the role of Project Blue Book to investigate unidentified aerial phenomena.

In 1985 the US Air Force stated there had never been extraterrestrials or equipment on Wright-Patterson Air Force Base.

In February 1993 the National Air and Space Intelligence Center at Wright-Patterson Air Force Base in Dayton, Ohio was designated the US Air Force unit for analyzing military intelligence on foreign air and space forces, weapons, and systems.

A 1994 Air Force Report revealed the Mogul balloon story was fake. The UFO spacecraft crashed in a field near Roswell.

The AATIP Advanced Threat Identification Program was shut down in 2012.

After 2012 the Pentagon in Washington DC took over investigations on sightings of UFOs and extraterrestrial alien spacecraft.

Declassified Extraterrestrial Alien Encounters and UFO Events

1947 July 7 Phoenix, Arizona William Rhodes took photos of an extraterrestrial gray disk flying saucer hovering about 5,000 feet.

1947 July 26 National Security Act of 1947 made extraterrestrial aliens and UFOs a Top-Secret classification.

1948 June Astrakhan Oblast, Soviet Union, Kapustin Yar the primary ballistic test site for nuclear missile development experienced an ET cigar-shaped UFO. The Soviets shot down ET bodies and technology.

1949 August 29 Kazakh Republic of the Soviet Union at a secret nuclear testing site where the second country to explode an atomic bomb named First Lightening came under the surveillance of UFOs.

1952 July 19 Washington, DC sighting of 5 UFOs.

1952 July 27 Washington, DC US jet pilot chases UFOs over the Capital.

1952 September 14-25 NATO-US Navy North Atlantic Sea Operation Mainbrace reported many UFO sightings during the exercise. Photographer aboard USS Roosevelt captured pictures. Sir Winston Churchill and the British military confirmed the UFO events.

Rudloe Manor at Wiltshire England was designated by the British Ministry of Defense as the facility for storing crashed UFOs.

The next War of the Worlds is more likely a war against ETs and advanced technology spacecraft than between countries on planet Earth.

This gives legitimate cause for secrets. A secret outer space program for defense of Earth must be inherently closed.

A secret space program exists with a different military philosophy. Technology and science lead its destiny.

Space 6
Governments and Militaries Worldwide ET-UFO Encounters

1953 January 14 Washington, DC the Central Intelligence Agency (CIA) reacted to the new rash of UFO sightings by forming a special study group within the Office of Scientific Intelligence (OSI) and the Office of Current Intelligence (OCI) to review the situation. During World War II the Office of Strategic Services (OSS) – the forerunner of the CIA – became aware of questionable events inside Germany giving rise to new technology.

The **Robertson Panel was formed in December and first met to investigate and study the phenomena.** The panel concluded unanimously that there was no evidence of a direct threat to US national security in the UFO sightings as a public announcement to avoid hysterical mass behavior. Nor could the panel find any evidence that the objects sighted might be of extraterrestrial origins.

Top Secret recommendation followed to debunk all UFO sightings and issued propaganda statements in a counterintelligence operation to divert public attention.

1953 November 23 Comox, British Columbia Canadian Forces Base Lake Superior US Air Defense Command witnessed a fast-moving UFO blip on radar collide with a Kinross Air Force Base jet. One pilot ejected and the body of the other pilot and both crafts were never recovered.

1954 January to September US Air Force investigated 254 UFO sightings. Conclusion was the UFO spacecraft were coming from Mars.

1954 February 20 Edwards Air Force Base President Dwight Eisenhower during three meetings signed an agreement with an extraterrestrial alien from an extraterrestrial civilization.

The treaty agreed extraterrestrial aliens and UFOs can stay in our skies for surveillance and the study of us. In return, we get all extraterrestrial alien and UFO technology. We further agreed not to disclose the existence of extraterrestrial aliens and UFOs. Ike believed there was life elsewhere in the universe. As we are aware now this agreement has been nullified as the pursuit of UFOs continued.

1954 Austrian Mountains France picture of UFO spacecraft.

1955 August 21 Hopkinsville, Kentucky Elmer Sutton reported having fought off gray aliens for 3-hours.

1957 November 2 at 11 pm Levelland in west Texas witnesses saw an egg-shaped cigar-shaped UFO that cut off the engines to their cars and the craft changed from orange to bluish-green color as it landed.

1957 October 16 New Mexico Holloman Air Development Center light was photographed when it hovered for 15 minutes.

1960 December Washington, DC Brookings Institution an American research and think tank group wrote a 186-page report on how the public would react in chaos if told the truth about ETs and UFOs. It was believed markets would collapse and religion would vanish.

1963 November 12 Washington, DC President John Kennedy requested in writing to the CIA for disclosure of information on UFOs.

1964 January 22 Herald News first public news article to declare UFOs very dangerous problem to planet Earth.

1964 September Big Sur, California Vandenberg Air Force Base test-launched an ICBM Intercontinental Ballistic Missile traveling over 1,100 mph accompanied by a disk-like UFO craft beside the ICBM in flight.

The UFO easily glided beside and while maneuvering all around the ICBM shooting light beams into the missile causing the missile to vary off course. The ICBM test failed.

Publicly the CIA claimed it never happened. However, actual film footage and records of the event proved it did. The report concluded this was either a warning or a threat. ETs can take control of any nuclear facility at any time.

1966 April 6 Navy Lt. CDR Brian Westin flying a Grumman A6 Intruder with CDR Bill Westerman testified to a UFO encounter.

1966 April 6 Melbourne, Victoria, Australia the Westfall UFO encounter. A grey saucer-shaped spacecraft landed in a nearby field witnessed by hundreds of students and teachers at two schools.

1967 March 24 Great Falls, Montana Malmstrom Air Force Base Nuclear Facility experienced a glowing UFO overhead simultaneously shutting down all nuclear silos and deactivating all nuclear warheads.

The ET UFO with no engine noise took each nuclear missile off-line. Each missile silo was an independent system. It did not damage the missiles when they were inexplicably turned off. It did not hurt anyone.

Continued ... 1967 March 24 Great Falls, Montana Malmstrom Air Force Base Nuclear Facility, ET aliens have landed on planet Earth, infiltrated nuclear missile sites, and sabotaged nuclear weapons. According to USAF Capt. Robert Salas and 120 military personnel

The message was clear ... get rid of the nuclear missiles. It was confirmed by the US Air Force the ET UFO craft was not from Earth.

1967 May 17 Moscow meeting of top Russian scientists confirmed UFO craft evidence and sightings.

1969 October Leavy, Georgia President Jimmy Carter witnessed a UFO.

1970 June Returning from Vietnam flying B-52 Stratofortress over the Pacific Ocean USAF Capt. James Boshears encounters multiple Tic Tac UFOs making 90-degree turns at over 6,000 knots or 6,905 mph.

1973 January Lauderhill, Florida Air Force Base held ET alien artifacts, pieces of UFOs, and a dead extraterrestrial alien in cold storage.

1973 February 19 Homestead Air Force Base reported President Richard Nixon and comedian Jackie Gleason viewed the body of the dead extraterrestrial alien Gray species in cold storage.

1975 November 5 at Apache-Sitgreaves National Forest in Arizona multiple corroborative witnesses saw Travis Walton approach a UFO and being abducted for five days.

1978 December Antofagasta, Chile largest UFO ever reported to 10,000 feet in length equal to 10 aircraft carriers.

1980 December 29 Piney Woods, Texas Betty Cash, Vickie Landrum and grandson Colby came upon a diamond-shaped extraterrestrial UFO hovering above the trees. They stopped the car and when investigating felt intense heat emitting from the spacecraft later resulting in radiation sickness and skin cancer.

1981 January 13 New York 81st US Air Force Combat Group were under observation by unexplained lights.

Russia and UFOs

1981 March 6-8 President Ronald Reagan discussed his 1974 September 19 incident with a UFO flying alongside Air Force One with William Casey of the CIA. Reagan included in the discussion his February extraterrestrial UFOs encounter.

From this conversation came an important 2005 document about extraterrestrial species on Earth.

President Reagan asked Russia Mikhail Gorbachev to help defend planet Earth against an extraterrestrial alien invasion.

President Ronald Reagan announced on 1983 March 23 the SDI Strategic Defense Initiative nicknamed Star Wars sought to protect the United States from attack. The primary objective and real reason for SDI was not made public.

The real objective was to find extraterrestrial aliens, UFOs and to build a military defense system against extraterrestrial alien species and UFO attacks.

1982 October 4 from 7:30 to 9:37 pm Usovo Nuclear Missile Base Ukraine Soviet Union several ET UFOs took control of an underground military nuclear missile bunker. There were no indications of hostility from the lights.

ET UFOs started the launch sequence and could not be shut down. Then the missile launch suddenly stopped before launch.

For a short time, signal lights on both control panels suddenly turned on, the lights showing that missiles were preparing for launching. This could only happen if an order were transmitted from Moscow testified Colonel Boris Sokolov.

ETs monitoring Earth demonstrated with sci-fi technology the taking control of a nuclear launch site putting all of Earth on notice. This message made clear to not use nuclear weapons. This open invitation maybe the last time we receive just a warning.

Major M. Davidovich Kataman Senior Assistant to the Commander of the Military Unit 52305 Service in charge of the computerized control panels for the long-range nuclear missiles at the Usovo Base was on duty in the underground bunker. Major Kataman did not see the ET UFOs flying above but did see the elements of nightmares when spontaneous illumination of all display indicators lit up on the panels.

The ET UFO had taken control of the nuclear codes and initiated manipulating code combinations of the computer-controlled launch panel.

Testing of apparatus and measurement of parameters according to technical Map 1-30 showed no defects. The apparatus was functioning normally before and after the event of the phenomenon.

The ET UFOs had their own reason to be there, their own program, and they could not care less whether or not Moscow liked it or not. They almost triggered a nuclear war testified Ufologist Paul Stonehill.

1984 Russian Academy of Sciences UFO Commission acknowledged the USOVO 1982 October 4 files confirming underwater UFO bases.

1984 July 12 Soyuz 7 Soviet Cosmonauts reported 7 extraterrestrials angelic like beings 80 feet tall with giant wing spans and glowing halos.

President of Russia Vladimir Putin is aware of the existence of ETs and UFOs and their importance to the space program.

Russia leads in many areas of outer space science. There are a couple of reasons for this. First is their education system which promotes the study of technology and the sciences starting at an early age. Second is their emphasis on the study of physics and advanced mathematics.

Though Russia is the largest-size country in the world, they have a smaller population compared to other more populated countries. To make up for this short-fall Russia has established its power-and-influence-base predicated on its military and intelligence capabilities which includes its investigations of ETs and UFOs.

1985 September 9 Mojave Desert British Pilots Picture of UFO

1986 January 29 Dalnegorsk Russia 7:55 pm UFO crashed. When the metal from the spacecraft was exposed to heat, it changed to other nanotechnology machinery metals. It has been referred to as the Russian Roswell.

1987 June 20 Arkaim Southern Ural Russia found evidence not of this Earth of advanced extraterrestrial civilizations and UFO spacecraft.

1989 March 25 UFO 50-miles long was identified outside Earth's atmosphere with an infrared image taken by a Russian Martian space probe Soyuz 2 approaching Mars as a Mother ship from a distant star.

1991 Moscow Soviet Union 7:00p President Mikhail Gorbachev just prior to his resignation released Russian UFO secret files.

1992 August 1 11:20p Gagetown Base Canada US Army David Marceau and others testified an UFO alien spaceship hovered for 30 minutes. It was oblong with rounded corners, an acre in length, and terrifying.

1995 Fall USAF Malmstrom Base Montana KC 135 pilot Aerospace Physiologist Maj. Daniel Gibson testified to witnessing an UFO hovering in the Aurora Borealis before shooting straight up at an acceleration of 3,700+ mph out of the atmosphere of earth.

1997 March 13 Phoenix, Arizona, Henderson, Nevada, and Mexico repeatedly viewed V-shape spacecrafts.

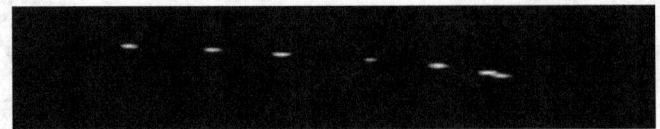

1997 March 24 Phoenix Lights former Arizona Governor Fife Symington testified to V-shaped UFO witnessed by thousands of residents.

1999 March 24 USAF Boom Operator Senior Airman Derek Tarr flying in an AC 10 over the Adriatic Sea encounters a circular-shaped UFO.

2000 Black Sea Turkey UFO sighting at a 7,500-year-old farmhouse. In Paracas, Peru an enormous, elongated nonhuman skull of extraterrestrial alien origin was found.

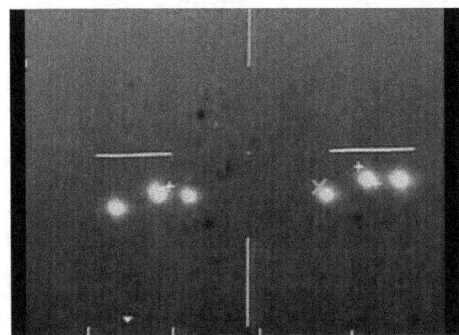

2004 March 5 Campeche Mexico Mexican Air Force Jet Pilots filmed UFO lights using infrared equipment at an altitude of 11,480 feet.

2004 November 10 Santa Catalina Island San Clemente Island US Air Force Base San Diego California USS Nimitz Lt. Chad Underwood F/A-18 Super Hornet film footage of Tic-Tac shaped UFO.

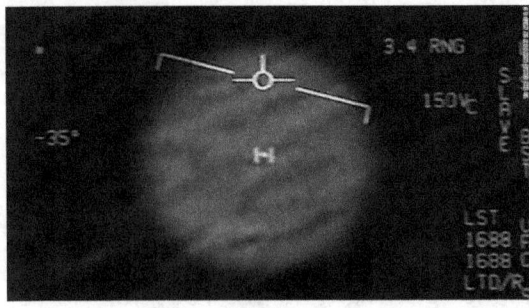 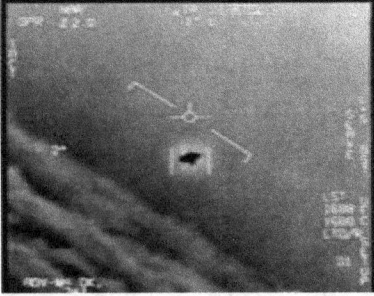

The outer space craft had no windshield, no wings, and no exhaust. The UFO picture was verified by the US Navy.

2004 November 14 USS Nimitz Carrier Strike Force Fleet in the middle of the day witnessed over 100 UFOs. The first identification of the UFO sightings was by a Northrop Grumman RQ-4 Global Hawk UAV.

2005 September Toronto Canadian Defense Minister Paul Hellyer acknowledged UFOs existed and planet Earth should be making preparations for an intergalactic war with extraterrestrial civilizations.

Hellyer shared there is a race of humanoid-esque beings called the Tall Whites living among us believed the closest species to the human race and possibly our origin. The Tall Whites ETs are working with the US Air Force in Nevada.

Governments of the world are continuously trying to cover up the fact that there known identified 80 different species of extraterrestrial aliens in existence on planet Earth.

The events and evidence of phenomenon pointing to ET and UFO experiences keep growing. Meanwhile we mere mortals are wondering how we fit into the picture. There appears a parallel to the advancement of our science and technology with an incredible dimension. We are knocking on the door and asking to be let in.

2007 the first FRB (Fast Radio Burst in radio astronomy) radio pulse signal was detected by Duncan Lorimer and David Narkevic looking through archival pulsar survey data by the Parkes Observatory on 2001 July 24. Hence the name Lorimer Burst.

2007 October 31 Philadelphia Pennsylvania during presidential debate presidential candidate Dennis Kucinich disclosed on one evening of September 1982 he saw 3 UFO spacecrafts.

2008 January 8 Stephenville Texas 50 witnesses to massive size UFOs.

2009 September 4 Hellen Province Afghanistan Army First Class Specialist Justin Doerfler sighted a UFO using night vision infrared measuring temperature differences rather than ambient light allowing objects as UFOs to be seen in complete darkness.

China and UFOs

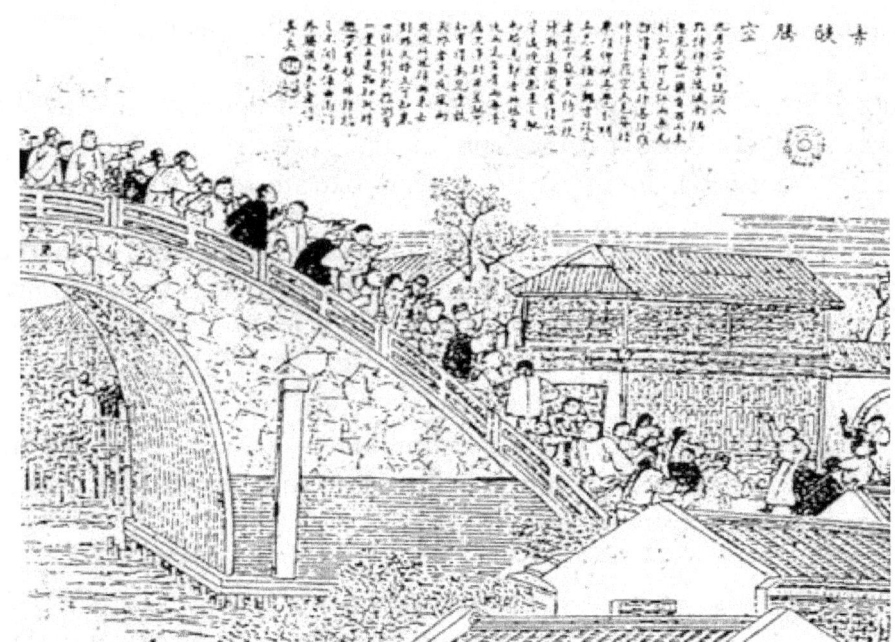

Chi Yan Teng Kong (赤焰騰空) Red Flame Take Off - Soaring in the Sky

The first graphical pictorial of UFO phenomenon in China occurred during the Qing Dynasty was painted by the artist Wu You-ru (吳友如). In 1979 Zha Leping at Wuhan University initiated the first UFO study called the Liaison Office for UFO Enthusiasts. In 1990 the study of UFOs were standardized with the Yiwu UFO Research Society. UFO sightings had increased to more than 100 reports per year.

China Close Encounter of the Third Kind
On 7 June 1994 Meng Zhaoquo while investigating a white shining phenomenon in the Red Flag Forest of Heilongjiang was abducted aboard a UFO spacecraft. July 17 Meng was abducted again at his home and shown Mars. The ET aliens explained this was their homeworld.

This close encounter of a third kind was investigated by the China UFO Society. The body of the society was composed of distinguished teachers, students, scientists, government officials, President Wang Changting, Chen Jingrun who solved the Goldbach Conjecture mathematical problem, Beijing Municipal Science and Technology Commission, and the Beijing Association for Science and Technology.

President of China Xi Jinping is aware there have been more frequent ET and UFO events in China. There are now over 5,000 reports per year of UFO phenomenon in China.

2010 July 7 Hangzhou China Hangzhou Xiaoshan International Airport closed about 9pm and flights rerouted when the UFO was detected. Government of China claimed the UFO was a Chinese military aircraft.

Original Photo of UFO Hovering Above Xiaoshan International Airport

Take note the UFO was hovering and still, not in motion. This was not a military flyby. Crews of two separate flights discovered and sighted the UFO. At the time it was disturbing enough for officials to take immediate action to learn what the UFO was after. Accounts of the incident was reported in several sources and then deleted.

On July 25, the BURO Beijing UFO Research Organization completed its joint investigation of the incident and concluded the phenomena is a photo of a private or military aircraft as reported in the Beijing Times and People's Daily. No evidence was obtained to show the UFO was associated with an extraterrestrial flying saucer according to General Zhou Xiaoqiang, Board Member Zhang Yunhua, Shanghai UFO Research Organization Director Lou Jinhong and Deputy-Director Bai Tao.

2010 January 26 London England Royal Society announced life exists in the universe extraterrestrial to planet Earth.

2010 September 27 Washington DC National Press Club 6 high-ranking military officers gave testimony of their verified encounters with UFOs. They are sending us a message ... was the central theme of the presentation. USAF Colonel Charles Halt, USAF Lieutenant Colonel Dwayne Arneson and USN Commander Patrick McDonough

2010 October 23 FE Warren USAF Base Wyoming Airmen in missile silos testified to witnessing over 100 ICBMs taken offline for 1-2 minutes by a hovering UFO over the base.

2011 January 28 Jerusalem Israel UFO hovered for hours above the Temple Mount House of God on a Friday Sabbath.

2011 The military of China spotted in the Gobi Desert runways over 1.15-mile long and 0.65-mile wide being used as a testing ground for extraterrestrial alien spacecraft.

China has the longest recorded civilization of astronomy from UFO to Dragon to First Emperor onto the times of the earliest emperors.

The search for extraterrestrial life continues with the interests of the Defense Minister of China, the People's Liberation Army (PLA) Rocket Force and the government of China.

The National Astronomical Observatories of Chinese Academy of Sciences (NAOC) operates the largest astronomy radio dish on Earth. The observatory is hidden in the Karst mountain region more than 100 miles away from Guiyang the capital of Guizhou.

The world's largest telescope is located in China's new Pingtang Astronomy Town. The radio dish is an advanced military defense radar and customized to search for extraterrestrial life and intelligence.

Fast Radio Burst Radio Telescopes and Military UFO Sightings

The Magellan Telescope

The anticipated 2028 completion of the Giant Magellan Telescope located in the Atacama Desert 71 miles northeast of La Serena Chile will be the world's largest optical telescope.

2012 December 7 Moscow Russia Dmitry Medvedev President of Russia revealed his knowledge of the existence of extraterrestrial alien visits and they reside on planet Earth. UFOs exist and defy explanation.

2013 May Washington DC Canadian National Defense Minister Paul Hellyer addressed the issue of ET UFOs more advanced than ours may be humbling.

There exists evidence nuclear weapons had been used in the past on Earth and other planets which destroyed their planets as a habitable place. We are being observed.

When you split an atom into for nuclear warfare you tear into other dimensions that cannot be allowed. Earth has to live in peace with our ET neighbors. Defense Minister Paul Hellyer

This testimony is told over and over in several of these events.

2013 September 18 Wittenberge Germany a whole fleet of UFOs were witnessed by the town.

2014 Summer to March 2015 US entire Atlantic east coast and Pacific west coast, Arabian Gulf daily whole fleets of UFOs stalking traveling at hypersonic speeds for months. Theodore Roosevelt and Nimitz carrier strike groups, squadron fighters, and USS Nimitz witnesses. Navy acknowledged UFOs are real and they do not belong to our civilization. USN Lt. Ryan Graves, USN Lt. Danny Aucoin
USMC Lt. Col. Chris Cooke, Lockheed Martin Skunk Works Steve Justice

Now we have the proof ... Sec. Defense for Intelligence Chris Mellon.

Comment
Every country on Earth may be in trouble. The outer space powers of the United States, Russia, and China know this very well.

As significant military powers, the need for advanced outer space vehicles is great. Especially of the military kind for extraterrestrial exploration and fighting extraterrestrial wars.

Life forms on Earth came from the stars and will return to the stars. Whether by purpose or accident it appears through permutations and mutations human evolution was manipulated.

Human genes could not have developed by chance or inclination. Investigations of ETs has raised an intriguing scientific question as to whether extraterrestrials exist or all Earthlings are human.

This is an important factor to know when developing a new intelligence agency for the extraterrestrial. What we understand at present may well change with extraterrestrial travel and exploration.

Earth's understanding of the extraterrestrial is in its infancy. It will only succumb to conclusion with a space intelligence agency to protect.

For planet Earth to keep its guard up is still a mystery of sorts. We will only progress through the unveiling of the unknown and progress alongside that which we lack understanding.

Chile and UFOs

2014 November 11 Santiago Atacama Desert Chile government and military confirmed UFOs using anti-gravity technology with centrifugal force for propulsion driving.

They further elaborated the UFO was generating its own gravitational spherical field suspending gravity around the craft like an Ezekiel wheel spinning on their own axis to make sudden accelerations.

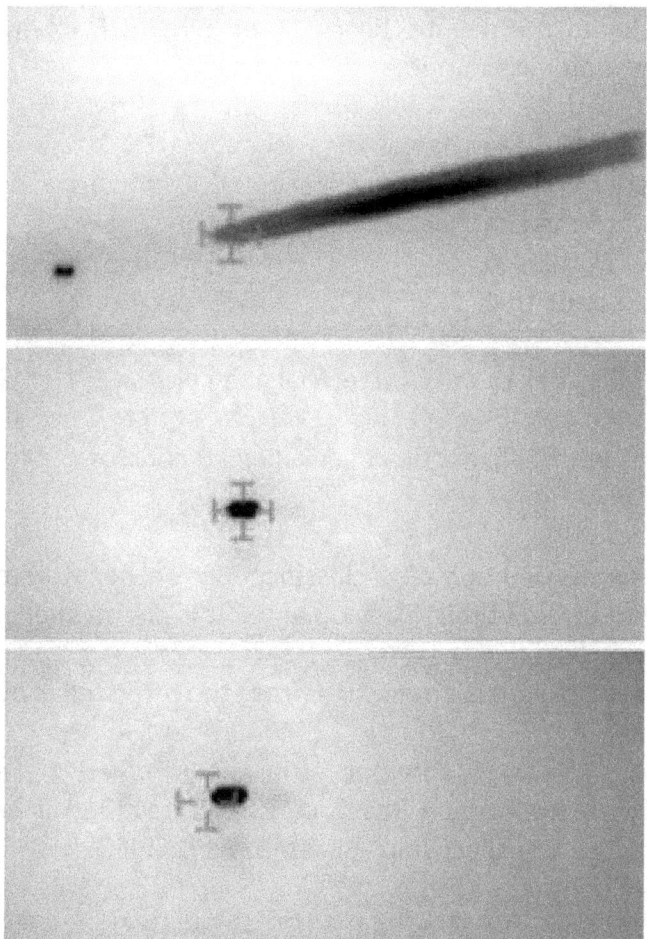

Three Separate Stills from the Chile Government Video Release

The Chile Navy video was filmed from an Airbus Cougar AS-5322 helicopter west of Santiago flying at 4,500 feet and 152 mph. The video shows a dark, disc-shaped unidentified object flying above the sea.

The Navy of Chile video showed two instances where there appeared an ejection of plume material discharge of some type of gas or liquid with a high thermal track or signal seen in the top still.

A 2-year investigation by the Committee for Studies of Anomalous Aerial Phenomena (CEFAA), a Chile government body designated to investigate Unidentified Aerial Phenomenon under the control of the Chilean Air Force concluded the UFO was real.

It concluded the naval officers filmed a real UFO flying over the ocean for around nine minutes.

2017 October 19 Maui Hawaii cigar-shaped UFO 2,625 feet or 800 meters in length recorded as traveling 195,000 mph around Earth then moving toward the sun.

Asteroid or ET UFO ... ?

Interstellar rock or extraterrestrial alien visitor ... ?

Passing interstellar object or alien technology exploring the cosmos ... ?

Astronomers do not have a classification for the interstellar object or agreement for what it is. They do agree it is not an alien spaceship.

There are no signals emanating from the interstellar object.

Andrew Siemion Director of the Berkeley Search for Extraterrestrial Intelligence Research Center testified it is a remote possibility that it is a spaceship – a remote possibility.

The US Pentagon in Washington DC secretive Advanced Aerospace Threat Identification Program is well aware and keeping an open eye on the situation.

2017 October Journal of Nature Astronomy report from the University of California Santa Cruz stated the discovered asteroid features of its shape, color, and tumbling motion was a product from its origin.

On November 21 the first proven extraterrestrial alien interstellar spacecraft verified entering our solar system was named Oumuamua.

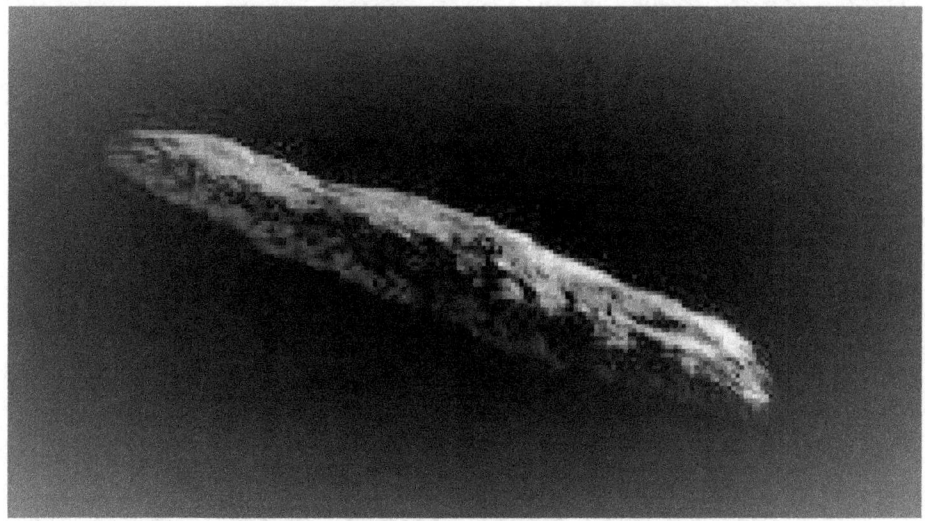

Oumuamua

Stephen Hawking referred to the elongated cigar-shape alien asteroid spaceship 1/4 mile-long and 260 feet wide traveling 196,000 mph as an alien probe.

2018 September 6 Sunspot New Mexico US National Observatory was shut down and evacuated due to the reporting of UFOs.

2018 September 16 Researchers with the Canadian Hydrogen Intensity Mapping Experiment Fast Radio Burst Project detected the first pattern in bursts occurring repetitiously.

Scientists were unsure what the patterns meant. While a dish can be aimed at specific targets, the CHIME radio telescope is stationary and has no moving parts. It monitors the heavens along with the movement of Earth.

2019 March 28 Romanian engineer Razan Sabie in conjunction with scientist Iosif Taposu invented the ADIFO All Directions Flying Object or first successful manmade flying saucer. This was accomplished by being the first to solve the aerodynamic all directional problem of flying saucers.

The research proved a gravity propelled craft is all technology and aerodynamics is irrelevant. The shape of the craft does not matter as much as the anti-gravity propulsion system driven by centrifugal force around its axis.

2019 April 23 Pentagon Washington DC admitted the UFO threat and required all UFO sightings of the military reported. On May 22 the Department of Defense admitted it investigates ETs and UFOs.

2019 June 2 China Bohai Naval Exercises UFO sighting. Foreign Ministry Official President of China UFO Research Center Beijing Branch believed ETs live among us. Chinese President XI Jinping called for efforts to develop China into a world leader in science and technology.

2019 July 1 Osaka Japan President Trump interview told Tucker Carlson he was not a believer but anything's possible and did not want to get into it too much. President Trump knows but cannot admit.

Government archive documents revealed 36% of all US Presidents had encounters with extraterrestrial UFOs from George Washington and Abe Lincoln to more current presidents.

The Alien Agreement of 1954 President Ike Eisenhower signed required no public admission of the existence of extraterrestrial aliens or UFOs. For years this silenced Presidents of the US as restricted to what they could say but more importantly to what they were told.

2019 August 29 UK Minister of Defense Nick Pope interview believed the information about UFOs and extraterrestrial aliens is so terrifying it would be contrary to the national interest to reveal it to the public.

Anytime there are nuclear tests there are ET UFO intelligence operations. Over 800 UFO sightings have occurred exactly duplicating surveillance at nuclear testing facilities. This was verified by eyewitness testimony from all branches of the military.

ET aliens have landed on planet Earth, infiltrated nuclear missile sites, and sabotaged nuclear weapons.

Every variety and shape of military ET spacecraft represents a different extraterrestrial civilization fighting over planet Earth. Each spacecraft coming from different planetary systems throughout the cosmos represent their own agenda and interest.

There are links between the shapes of ET UFOs and the types of ET entity beings.

Competition of world powers currently exists to manufacture military technology advanced spacecraft based on ET UFO exotic design. UK Minister of Defense Nick Pope

2019 August 30 a second discovered elongated cigar-shaped alien asteroid 21/Borisov was identified having 66 antennas.

Illustration of What Astronomers Witnessed

2019 August 27 Boca Chica Dominican Republic SpaceX-base preparing the Star Hopper spacecraft for transporting human beings to Mars.

2019 September 25 Pasadena California NASA Jet Propulsion California Institute of Technology Laboratory Center witnessed artificial intelligence shapeshifting probes believed designed to explore planets.

2019 October 30 Using a CHIME telescope, researchers with the Canadian Hydrogen Intensity Mapping Experiment Fast Radio Burst Project detected a repeating pattern of recurring signals releasing a radio burst or two each hour and then went silent for 12 days.

2020 January astronomers reported the precise location of a second repeating FRB. In February astronomers detected FRB (Fast Radio Burst in radio astronomy) radio pulse signals to repeat in a regular way every 16.35 days emitted from a galaxy 3 billion light-years from planet Earth outside our Milky Way of extraterrestrial intelligence.

Many scientists in order to protect their careers have ruled out aliens. As more FRBs are discovered will be the only way to know for sure. This may be a pulse dispersion from an exploding star. It may turn out some FRBs are transmitted signal communications from an unidentified source. Whichever the case or both, time will tell.

Interestingly, North America UFO reports increased from 3,395 UFO sightings in 2018 to 5,971 UFO sightings in 2019. The top rankings for the most UFO reported sightings were California, Florida and Washington. National UFO Reporting Center

2020 January 27 at 11:18 before midnight in Mexico a fast-moving UFO was seen near Popocatepetl volcano when it exploded.

Italy and UFOs
The government of Italy acknowledged the first attack by a UFO on a military aircraft. The extraterrestrial incidence was significant having military implications for all countries and militaries on planet Earth.

In 2004 an Italian military helicopter in Sicily was filmed being shot by a laser ray from a UFO. The sci-fi technology was real and the spacecraft was of extraterrestrial in origin and not from planet Earth.

ET UFO Military Capability
The spacecraft used a powerful electromagnetic ray-beam in a certain band of frequencies. The energy weapon destroyed the Italian Navy helicopter rotor on the wing as it flew over the water on the coast of Sicily. The source of the ray-beam was 800 kilometers in the open sea from the Italian military helicopter as it hovered.

Missiles could not be used to defend or attack the UFO that was seen as a bright light underneath the surface of the water. The UFO energy weapon ray-beam displayed the capability to capture a missile and return it back to its origin place of launch.

No military on planet Earth has the technology to do this.

How to Destroy an Enemy Military ET UFO
Lessons were learned by the Italian military how to defeat and track a military UFO enemy. We should use the same frequency radio-beacon to track ET UFOs and show them we understand their technology.

An ET UFO must go outside its stealth mode in order to attack with energy weapons. While the UFO is exposed and vulnerable to attack, the militaries of the Earth can then attack with uranium grade weapons during this time when the ET UFO is not cloaked in a defense shield.

The Centro Ufologico Nazionale of Italy has logged over 13,000 UAP or UFO cases. This is a global problem and threat calling for a united planet Earth such as your author's Space Federation Council. Scientist Roberto Pinotti President of CUN, Italian Navy CDR. Clarbruno Vedruccio, Cpt. Bruno Cocciolo, and Tom Delonge Founder To the Stars

Australia and UFOs

2020 May 22 Victoria southeast Australia slow horizontal-moving UFO spaceship believed headed to its base was witnessed by over a thousand persons. Space scientists ruled it could not have been a meteor since they typically last three seconds at most.

The Australian government announcement claimed the UFO phenomena was space junk descending from orbit. The fact the UFO was slow-moving at a shallow angle and an amount of disintegration was occurring disputes it was an alien spacecraft, meteor or comet. Astronomer Perry Vlahos President Mount Burnett Observatory

2020 July 7 Boolardy western Australia Astronomers confirmed findings of four mysterious blasts of cosmic radio waves at the CSIRO Murchison Radio-astronomy Observatory using the ASKAP Murchison Widefield Array Pathfinder telescope.

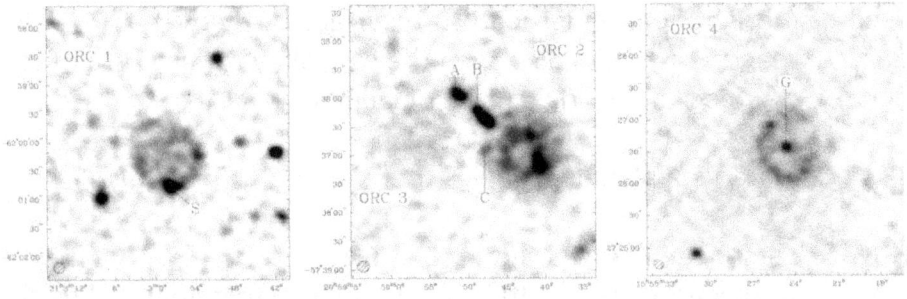

The circular glow is a new type of brightly-edged ring-shaped phenomenon only visible in radio wavelengths. They are invisible in X-ray, optical and infrared wavelengths.

They are not as yet understood since their signatures from outside the Milky Way do not match any known planetary nebulae or profile of supernova. Astrophysicist Ray Norris and his team have identified six other but fainter blasts of cosmic radio waves.

ET and UFO beliefs continue to have a growing acceptance around the world. Despite minor differences, witnesses continue to come forward worldwide sharing similar encounters with UFOs. While the world waits, the danger continues to grow.

Space 7
Space Forces

Russia Space Force

1992 August 10 Moscow Russia establishes the first Space Force as an independent non-military section within the Russian Armed Forces.

1997 July the Russia Space Force was incorporated into the Strategic Missile Forces.

2001 June 1 Russia Space Force was established. It became once again an independent non-military section of the Russian military.

2011 December the Russian Space Force and the Russian Aerospace Defense Forces merged into one military service.

2015 August 1 the Russian Aerospace Defense Forces and the Russian Air Force merged to form the Russian Aerospace Forces. It became one of the three sub-military branches and no longer an independent military entity.

Russia Space Force Badge   Russia Space Force Flag

1955 June 2 Russia space exploration began with Baikonur Cosmodrome the first and largest operational spaceport in the world located in southern Kazakhstan. In 1971 this space base facility launched Salyut 1 the first space station.

Russia Space Force performs a wide range of military and scientific missions. Space exploration and identification of potential threats are important to the Russian agenda as well as investigating and monitoring ETs extraterrestrials and UFOs.

China Space Force

2014 October the China Space Force was established as a fifth military branch of the PLA People's Liberation Army under the Aerospace Office of China and reporting to the CMC Central Military Commission.

 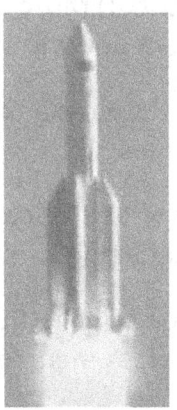

China Space Force Military Patch   China Deep-Spacecraft

2015 December 31 the China Space Force became the People's Liberation Army official Rocket Force Strategic Support Force (SSF) or PLASSF for space exploration and aerospace operations.

China Space Force Emblem   China Space Force Badge

1970 April China launched its first satellite.

1993 April 22 the CNSA China National Space Administration was established.

2019 January 3 China landed a probe craft on the far side of the Moon.

China has a rapidly advancing military space program in place to launch a Mars explorer in 2020 and complete an Earth-orbiting space station by 2022. China is looking to make the first contact with alien ETs in the extraterrestrial.

United States Space Force

2019 December 20 Washington DC United States Space Force (USSF) was established as the sixth branch of the United States armed forces with the signing of S.1790 National Defense Act for Fiscal Year 2020. This should not be confused with the AFSPC Air Force Space Command activated on 1 September 1982 at Peterson Air Force Base in Colorado.

2020 January 24 Washington DC United States Space Force (USSF) was activated as a separate military branch to protect and defend the national sovereignty of the United States and interests of the United States in outer space and conduct extraterrestrial operations.

US Space Force Badge

US Space Force Flag

The first mission will be a technology-focused military space force. It will require realignment and reorganization in the first few years as the USSF develops skills and assets of the most advanced technology.

I think it will begin with limited rules of engagement and believe its primary role will be to prevent war in space. This is all wrong.

First there will be a need for highly trained and equipped **Outer Space Military Troops** to fight wars in extraterrestrial environments.

Second there will need to be established in the cosmos **Extraterrestrial Military Outposts** as travel and exploration expands.

Third there will be a need for the most critical component of the US Military, a secret intelligence outer space service for extraterrestrial infiltration, intelligence collection, spycraft, espionage, and war ... Headquartered on the Moon ... **The Space Intelligence Agency.**

Japan Space Force

2020 May 18 the JASDF Japan Air Self-Defense Force was established as a Space Operations Squadron reporting to Japan Air Defense Command Headquarters at the Yokota Air Base in Fussa.

The JASDF Space Domain Unit and the JAXA Japan Aerospace Exploration Agency will liaise in cooperation with the US Space Command and US Space Force.

Japan has joined the United States as a joint space force operated by the JASDF Japan Air Self-Defense Force at the Fuchu Air Base west of Tokyo.

Japan Space Domain Unit will be part of the Japanese Air Force and a component of Japan's Air Self-defense Force.

Japan Defense Minister reports Japan military pilots have never encountered UFOs.

It is Japan's official policy to deny the existence of ETs and UFOs.

FVEY Five Eyes
FVEY Five Eyes Intelligence of Space is the world's oldest military and economic consortium sharing classified information and cooperation which extends to the space defense of enemy ETs and UFOs.

The intelligence alliance includes the United States, Canada, United Kingdom, Australia and New Zealand. The Five Eye alliance is furthermore in association with Japan, France, Germany, and Israel.

1975 May 31 an intergovernmental organization of 22 EU European Union member states established and headquartered in Paris France the ESA European Space Agency association to promote space exploration for research and advancement of technology. Europe has explored the idea of a European Space Force.

2019 July 13 President Emmanuel Macron announced the creation of the France Space Force. The Space Command of France is independent from the EU European Union space program.

The French Space Force operates as a branch of the French Air Forces under the Space Command in Toulouse, France. France has decided to separate and become an independent space agency.

From Rocket Man to Spaceman and from Space Cadet to Space Force the language of the countries of planet Earth are viewing the extraterrestrial as an extension of military, political and economic power.

It was important to include the current assessment of separate and independent space forces to show where the world is headed in the foreseeable future. The buildup of new weapon systems seems to replicate with a more powerful capability of destruction.

It is promising the destructive power of advanced weaponry will be used to defend Earth instead of its destruction for political gain.

Whatever new course and perspective the space age brings should keep us all wondering if we have opened a Pandora's box.

Space Force Invasion

Space Force countries on planet Earth want to be there first. Whether establishing a Moon-base, space colony, or military outpost on another planet, it is a race to the stars to claim I was here first.

This is of strategic importance were an extraterrestrial war to occur. Having eyes above and weapons hidden gives a battle advantage.

How can we expect to have acceptable rules of occupancy on other planets when we have not settled our differences on Earth?

We cannot. We should not expect anything to change from the current condition.

So we might ask ourselves if it were ever possible to come together to form one Space Federation Council for planet Earth?

The answer is not now but later.

Space 8
Science and the Unexplained Reality of ETs and UFOs

Analysis Inconclusive

I have only reported on what has been Declassified and disclosed by governments and militaries.

This list is not exhaustive but a notable one. Looking over the dates, times, and events with so many instances of claimed contact and sightings makes one wonder how so many could be wrong.

Yet every one of the occurrences is given some rational explanation of why they cannot be. Disclaimers abound and rumors circulate. It leads one to an inconclusive analysis.

Only the few who claim first, second, and third of kind encounters believe their own stories. It is easy to dispel each story with someone who has never experienced a phenomenon.

The whole idea of UFOs and extraterrestrial aliens visiting Earth from other planets makes for great science fiction. Today's science fiction many times becomes tomorrow's nonfiction.

Whatever you think will decide your opinion ... if it is real or fake.

Preponderance of Evidence
The search continues for existing extraterrestrial life and intelligence.

We seem obsessed with trying to find intelligent life on other planets while we question if it exists on planet Earth.

ET and UFO sightings and experiences are often derided as a religious type of mysticism. Fanaticism has entered the picture as a label distributed throughout society on undeserving witnesses.

Even in the scientific community, the innocent are marked as a cult of UFO enthusiasts with weird imaginations. Their careers are threatened by their peers. They are shunned by their colleagues. This unnecessarily makes for false allegations and biased assumptions.

An unwarranted stigma is attached to those who have experienced an ET or UFO phenomenon. There has always existed in our history times of regrets and punishment of those whose actions and thoughts are outside of accepted convention.

The reason for this claim is those in government and the military or intelligence business say otherwise ... ETs and UFOs exist. Simply because you do not have the physical evidence in front of you does not necessarily mean something does not exist.

UFOs exist. To deny is a catastrophe.
Select Committee on Intelligence Senate Majority Leader Harry Reid

UFOs are real. UFO events are real. UFO sightings are real.
ETs reside on planet earth. ETs are real. Not knowing changes everything. Knowing changes everything even more.

With a reasonable amount of proof and physical evidence in-hand, governments like to keep things to themselves for military advantage in case of a war on Earth between countries.

But what if there existed a preponderance of evidence of a greater threat an extraterrestrial war is likely from outer space?

Past ... Present ... Future for Outer Space

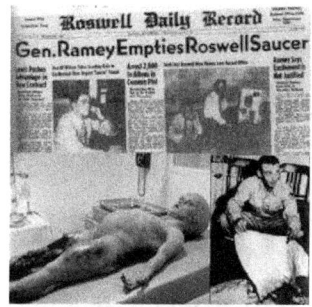  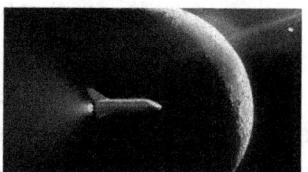

      Past   Present   Future

The belief we are not alone in the universe makes common sense. It is appealing to scientific studies and evidence of our existence. Science is without prejudice and emulates it is what it is.

We are still dealing with misunderstood technology and a world full of unexplained mysteries. Evolution and technology are the only things different between humans and nonhumans.

Some countries are starting to formalize a space force as an official branch of their military. They seem to view its mission as limited to signals intelligence gathering. It is not.

There is more at stake than just a country's survival or dominance over another country. The game-changer has spoken. The game-changers are science, sci-fi technology, and advanced intelligence.

Never in the existence of planet Earth is there a more compelling reason to prepare for the change that is upon us. It is another dimension of why now the world needs a Space Intelligence Agency.

There are plenty of skeptical nonbelievers. Rightly so because what we are talking about seems quite remote and unsupported.

People do not want to admit or believe even though ...

We are all offspring of star people.

Myths Symbols Facts
You will consider what you believe as pure fiction or authentic fact. Many times fiction is truth not yet revealed and proven.

We are in an era where our myths are becoming challenged facts. Symbols representing what beyond is known are now becoming visible.

The information presented in this book provides a lead into my second book calling for a Space Intelligence Agency. With the foundation established for the need of an extraterrestrial intelligence service and agency, it makes a timely case for critical action.

As related to the subject matter and importance of the extraterrestrial, we find ourselves involved on two fronts of contention.

First ... it is the sole intent of the author to grab your attention and stir your imagination as to what is or of what you believe is not, what might or might not be possible in the future for outer space interstellar travel and extraterrestrial mining and exploration.

This is the business side of the equation. The extraterrestrial is the greatest money-making invitation which we will ever have.

Second ... governments seem to have growing apprehensions about the extraterrestrial and its threats posed both outside and inside planet Earth.

This is the security side of the equation. The extraterrestrial poses the greatest threat to Earthlings and the planet in the future.

All written is claimed to be true. Though from the most trustworthy of secondary sources, I hope to create more interest in the future of outer space scientific research and travel.

What is not known, may someday be known. What was once impossible, the future may make possible.

There is truth in everything I have written in this book.

US & Earth Military Assets and Forces at Risk from ETs & UFOs

2020 June 24 in Washington DC the US Intelligence Committee announced approval requiring the Pentagon and intelligence communities to assimilate all data and evidence collected on ET and UFO phenomena consolidated into a declassified public report.

This is the first instance ever when a country asked the public to accept ET and UFO revelations as existing and admit the circumstances are beyond its defense capabilities to protect its population and planet. At this time there are no absolute defenses to the extraterrestrial.

This poses new rules on the US DOD Department of Defense. This placed the Department of Defense in a precarious situation.

The reason is they are scared and should be as you will further learn. Admissions of tracking and writing reports are not the same as saying we don't know.

The US Congress is trying to downplay the threat to the public by wording UFOs Unidentified Flying Objects as a term only referring to phenomena objects that cannot be explained and beyond our intelligence to comprehend. The intent is to discredit alien spacecraft.

By removing the term alien spacecraft from the definition of UFO demands its nonexistence. Governments find power in knowing and denying. What governments cannot explain they make SCI Sensitive Classified Intelligence. The burden of proof and evidence required is deniability.

When the very existence of governments are under threat of control by extraterrestrial powers and their militaries lack effective weapons to defend the sovereignty of a nation and the planet, you know there is a problem of some magnitude.

This real threat is posed to US military assets and installations as well as to all other military centers on Earth. They want to keep it top secret and out of the hands of politicians, enemies, and the public for good reason.

Intrusions in restricted air space by UFOs are something no military in the world is prepared to address, much less defend against. Simply wanting to firm up intelligence collection and sharing the subject of ETs and UFOs is an admission our intelligence capabilities are lacking. Thus another argument for the SIA Space Intelligence Agency.

It is my opinion the directive to compile data on unidentified aerial phenomena will still have a significant component of intelligence that will remain classified and far away from public consumption. Declassified will continue to mean partial intelligence as it should be under the present circumstances.

The reason for continued secrets is we are dealing with unknown much less understood quantities. Let's take a moment to review some of what we have learned so far.

To recap, we have heard the admissions on the existence of ETs and UFOs from heads-of-states of the most powerful countries on Earth, ministers of defense in several countries, military officers and troops, public witnesses having videos and military videos of like same.

The whole point is we don't need the government to tell us what we already suspect and know to be true to make it so.

We Earthlings are so silly to think that someone else always has to give us permission to accept something. What we think and believe never needs confirmation from someone else or another for us to accept our own thoughts.

The United States is the first nation to admit we are in trouble from an extraterrestrial threat. What governments all over the world do not want to acknowledge is there are people capable of figuring out what is going on and seek to determine why.

All countries have some very smart people. It may take some time to figure out the extent of our new consciousness. I believe it will happen sooner rather than later and Earth will have a solid fighting chance of survival.

It is interesting to also learn the US Intelligence Committee directed the reporting federal agencies to further explain how their data and information are collected and shared.

However, it was most interesting to further learn it would also require federal agencies reports to disclose whether particular ET and UFO incidents pose aerospace and weapons capabilities putting the US and its military forces at risk.

The intelligence gathered regarding ETs and UFOs is only for government consumption. Probably will remain so but everything eventually changes.

It appears we might have reached an apex when the evidence of the existence of ETs and UFOs is so great, like many events of the past, it is time to make some disclosure to the public without creating panic or feeling of helplessness.

This legitimizes the ET-UFO issue. While secrets remain secret it makes nonreporting a nonissue.

It is also my opinion the greatest danger at present is the reluctance of the public population to understand. This is mostly derived from a lack of education. Education to understand the science, use the technology, and rationalize the reality.

This same discussion would just as easily apply to every country and its population in the world.

Things happen for a reason. I am well aware more than most of the controversies surrounding such rhetoric.

With that said I see no alternative which would leave everyone with acceptance as to any one conclusion. It makes one wonder if we are truly in control of our lives or if our lives are more influenced by events than circumstances. Whichever the case, we will continue to live the life we have wherever we are.

Intelligence Agencies Prepare for the Extraterrestrial

The CIA, MI5, FSB, MSS, MOSSAD, every intelligence service and agency in the world need an update for the future in the extraterrestrial.

So like other countries ... the United States found it necessary to establish a military space force. The mission objective is limited to supporting the interests of the US and monitoring outer space technology.

Other countries on planet Earth are also following suit in the acquisition race to claim extraterrestrial territories on the Moon and other planets. In a short time, all this will change.

The United States and all the countries of the world have it wrong. They think they are creating an only country-military space force. The mission is all wrong. The structure is wrong. The outcome will be wrong.

In the past, this type of reckless thinking resulted in the loss of life and untold destruction. Revolutionary approaches at the beginning of now will better anticipate the world of tomorrow.

This discussion on this matter is the subject of my second book to follow about spies and espionage in outer space. I thought it more important in this book to also lay the groundwork for outer space travel, mining, exploration, and colonization in my coming third book about business by providing convincing data and information about the world of the extraterrestrial.

To put forward the idea of the need for an intelligence operation and apparatus before it is needed begs the question as to why.

As I see it all countries are in danger to the consciousness of the idea in the admission and existence of ETs and UFOs.

The good news is there is a way to guarantee a tomorrow ...
it may come from a collective spirit and a vision for the

Space Intelligence Agency.

Earth Is In Danger

Extraterrestrial alien ETs may be in control of planet Earth. Life is still far from normal. Unfortunately, it is not so clear what is normal.

Cases of ETs and UFOs have steadily been rising. Once this reaches critical mass fully containing it will be impossible.

Governments think the key here is slowly spreading through the population the existence of ETs and UFOs. Even if this is coming from above its going to be the reality on the ground.

Change is inevitable. Change is in motion. Change is in what we do not know.

The greatest weakness to planet Earth would be allowing any one country to aspire to independence of superiority both militarily and strategically were an extraterrestrial phenomenon event to occur.

No one country is capable of defending planet Earth were there to be an attack from the extraterrestrial. It would be of such magnitude the devastation would be complete.

The greatest Void of planet Earth is the absence of a non-identified enemy capable of seeking control of planet Earth.
Solution ... admit what we don't know and what we know is fact.

The greatest Vulnerability of planet Earth is the absence of intelligence of what is happening from outside of planet Earth.
Solution ... the SIA Space Intelligence Agency.

Planet Earth facing unseen deployments and energy weapons is vulnerable. It would be a mistake for us to turn them on each other. Science and technology have become the game-changer.

Book three of the outer space intelligence three-part series discusses how the infrastructure of outer space will be built upon a network system to support resupply lines and energy recharging stations. The extent of expansions will be determined by its resistance.

There is no economic advantage to what you don't control. Lack of a unified commitment in outer space for both commercial and military applications has a great potential to bring ruin, not extinction.

Looking for a total military and commercial advantage may initially not bring the returns desired in association with the costs. However, getting started is the most important part of the equation at this time.

This is best brought about through strategic cooperation since the competitiveness of Earthlings will never abate. With confidence, we can surely anticipate this is mild compared to extraterrestrial competition.

In this regard, a nuclear war causing a nuclear winter will serve to no advantage in any country. It will only make us as a planet more vulnerable and weaker to extraterrestrial hostilities.

Neither can we afford the outcome regardless of the military strategy to win. Populations want peace. Militaries want wars they can win. Governments want from wars conquests to expand their power.

What if there were an enemy of Earth greater than any one country?

It is suspicious that major countries are already using acquired foreign outer space technology to develop futuristic travel vehicles and advanced weapon systems. Arrogance is potentially dangerous.

Outer space entrepreneurs and business pioneers are right to take the lead in the search for and acquisition of new and familiar raw materials existing on interstellar planets. Governments are wise to support such endeavors. Populations may well come to depend on it in the future for their very survival on planet Earth.

In the larger picture space mining and exploration comes down to Earth in all forms of renewed raw resources. It all is about making the most money and accumulation of greater wealth. Grabbing control of new resources may create the future power base of a new generation.

Does the question then become who benefits?

Advantage to Being First in Outer Space
Being first in space is imperative for outer space power as countries fight to claim rich deposits of essential raw materials, expand their military space force and extraterrestrial intelligence gathering, and expand an interplanetary commercial free enterprise.

On 12 April 1961, the Soviet Union launched Yuri Gagarin, the first human and astronaut, into outer space to complete the first orbit around Earth. On 5 May 1961, the United States launched Alan Shepard beyond the atmosphere of Earth.

To develop the kind of future with safety and security desired a new commitment of ideas and a lot of money is necessary.

There is no longer time for increments of slow and cautious approaches. Immediate planning to climb the next realm of outer space awaits.

The greatest existing threat to planet Earth has been identified. ETs and UFOs indirectly by association and directly by contact having effective space capabilities is a call to arms.

This poses new allies and new threats. This means some extraterrestrials will be declared enemies and some will become enemies.

There is a great advantage to being first. Being in the lead gives control to the future of events and to not is to lose all advantage. The answer to the advantage of why being first in outer space is synergy.

The early years of outer space exploration were crucial to our emotional attachment of the future, development of skills for adaptation in the extraterrestrial, and improve innovative strength-based programs to improve our relationships with our neighbors.

What is sought is more than materials and power. The essence of all of this is how to fit into the scheme of extraterrestrial life in the cosmos.

Outer Space Industry a Quadrillion Dollar Market
To start, the marketplace of outer space is a multitrillion to quadrillion dollar valued market. Any entrepreneur would want to be the richest person in the universe. Imagine owning the cosmos.

Let's look at it this way. A country, any country on planet Earth, would want to rule Earth and the cosmos. It would want to control all finds of essential raw materials, new energy sources, and technology.

The urgency to seek and find whatever is out there is appealing to the senses and pocket. One cannot help but think about glory and riches. Or is it a Pandora's Box and military defense nightmare?

Solely designing country-to-fight-country defense capabilities and new weaponry may become obsolete for outer space security and defense before implementation. It may prove not only destructive and yet also a waste of money but a misdirected utilization of resources.

Any participation in a nuclear war on planet Earth jeopardizes the evolution of Earthlings and delays our participation to engage in outer space. In this instance, enemies bring enemies closer together.

Relying on yesterday's thinking will not help. Ignoring the thinking of what will not be admitted will not save us from an undefinable fate.

The unrestricted freedom of outer space private enterprise is to organize and operate for profit in the extraterrestrial expanse without interference from any government on Earth.

This makes for a paradigm shift in the power of governments and the operations of the military structure and response. To do this it is necessary to gather intelligence on all forms of ET extraterrestrials and infiltrate and perform all expectations of a spy agency intergalactically.

More than likely it is only a matter of time before planet Earth will square-off with ETs in a war. A war we are not prepared for or expect to happen. It is a dangerous proposition.

Space 9
Neighbors ... Moon Extraterrestrial Spy Base

The Moon is 238,855 miles or 384,400 kilometers average distance from Earth. It orbits the Earth every 27.322 days. NASA

The Moon has significant effects on Earth. It was created to make life possible on Earth.

Though about 25% the size of Earth, its equator is 2,160 miles or 3,476 kilometers in diameter. At the time of this writing, it takes about 8-to-12-hours to get to the Moon.

The Moon is always in synchronous rotation with planet Earth. It shows the same 59% side of the total lunar surface view from Earth.

Moon eclipses have guided timekeepers for thousands of years. Its influence on tides of the oceans and the four seasons have been chartered in many time periods. On 26 July 1609 Englishman Thomas Harriot made the first drawing of the Moon by using a telescope.

Galileo Galilei was the first to make calculations and record observations of the Moon and our Milky Way solar system. In July 1609 Galileo gave birth to astronomy by using his own invention of the telescope to see the night sky in some detail and all its splendor.

We see the same face of the Moon every night from Earth. This is because the Moon rotates around on its own axis in exactly the same time it takes to orbit Earth. The side of the Moon facing away from Earth, referred to as the dark side, cannot be seen from Earth.

The Moon does not produce its own light. It reflects the rays of the sun, maintains a fixed-orbit, creates the tides, makes the oceans stable, and keeps the axis of the Earth in position.

The sun shines equally on both sides of the Moon. The only times the dark side of the Moon has been seen by Earthlings is from passing satellites and spacecraft.

The Moon has no oxygen and thought at first no water. The surface of the Moon on its sunny side is 273 degrees Fahrenheit or 134 degrees Centigrade. In the shade on the dark side of the Moon, the temperature is a minus 243 degrees F or minus 153 degrees C.

The original theory of the Moon was created when it was thought a rock in space maybe about the size of Mars hit Earth some 4.5-billion years ago making the Moon the only natural satellite of Earth. Debris from the collision was believed to have clumped together to make the Moon. Scientists aren't in agreement with how the Moon formed.

However, in 1999 scientists discovered a 3-mile or 5-kilometer-wide asteroid caught in the gravitational grip of Earth thereby becoming a natural satellite. In February 2020, a small 6-12-foot diameter asteroid was discovered orbiting Earth and likely had been for the past three years.

The surface of the Moon was discovered to have mountains, huge craters, and flat planes called seas made of hardened lava. Like Earth, the Moon has gravity.

Want to lose weight? The gravity of the Moon is weaker. The Moon is only one-sixth of the gravity of Earth. That means you would weigh much less if you were to stand on the Moon.

Of significant importance is the relationship of the Moon to the existence of Earth. Without the stabilizing factors of the Moon, life would not be possible on Earth. The moon is like a guardian angel.

The Moon is not round or spherical. It is shaped more like an egg. The Moon has a solid metallic surface. Therefore we conclude, what is seen as scars in the craters on the Moon have little to no deterioration and have not eroded much for two main reasons ...

First ... the Moon lacks the type of geological activity we experience on Earth from Earthquakes, volcanoes, mountain-building, and sink-holes. The landscape on the Moon stays pretty much intact.

Second ... the Moon has virtually no atmosphere as there is no wind or rain to cause surface erosion. Nearly the entire Moon is covered by a rubble of charcoal-gray powdery dust and rocky-debris.

The light and dark areas represent sedimentary rocks of various compositions and ages. This is evidence for how the surface-crust of the Moon may have crystallized and preserved the craters.

We have little verifiable proof and little sound evidence of the facts we keep repeating. It might be better to start all over and consider what we now do know.

The Moon is of the perfect size and placement positioned as the exact size of the sun and the exact size to eclipse the sun. The Moon is in perfect-pitch orbit to protect planet Earth from radiation sun bursts.

Our best science and engineering scientific principles argue it is not possible the Moon came about through natural evolution. Thus it was not by accident but by design. Let's take a close look at some of the factors which make it so.

Scientific evidence says the Moon is hollow with impact craters from space debris. This is a theory not widely believed to be true. There is no presence of natural erosive forces and all craters are of the same uniform depth.

It is a wonder the Moon even exists. We don't know who built it. How it was built. When it was built. Where it was first observed. In fact, we are still in the dark about much of our extraterrestrial environment.

In 2009 we learned there are water molecules in ice on the Moon. There are iron and nickel below the surface in the south pole basin. The South Pole-Aitken basin is the largest impact crater in the solar system.

Discovered by seismic equipment on the Moon from the Apollo missions 1969 to 1977, the Moon is minutely shrinking as its interior crust continues to cool down. This resulted in Moonquakes from faults and cliffs movements, not from previously thought asteroid impacts or energy activity deep inside the core of the Moon.

Apollo astronauts used seismometers to discover the gray-orb of the Moon is an active place in a geological sense. Space-age exploration continues to show us how the Moon is connected to human existence.

The Moon may not make you rich. Yet it may become your launch-pad into the extraterrestrial for travel, exploration, and space mining.

You won't strike it rich on the Moon unless you know what is inside its core deep below the surface. This is where space mining comes into play in my third book in the outer space intelligence series.

First Contact

The first spacecraft from Earth to reach the Moon was Luna 1 from Russia in 1959 and Luna 2 was the first spacecraft to impact the Moon. In 1966 the first soft landing on the Moon was achieved by Luna 9.

1969 July 20 Apollo 11 from the United States was the first spacecraft to land humans on the Moon. Neil Armstrong was the first Earthling to walk on the Moon. The Apollo 11 lunar module *Eagle* was piloted by Buzz Aldrin.

300 kilometers away from the Apollo 11 landing were several obelisks as found in ancient Egypt on Earth. Pictures of the obelisks on the Moon were transmitted to the NASA center during a two-minute cut of the broadcast loss of transmission to the public.

The astronauts had a private medical channel for security to talk to NASA. This secure link was used to verify what they saw which changed everything we thought we knew beyond Earth. The conclusion from the findings was the Moon is of superior intelligence-design and built as a space station. It leads to the question as to how the SIA Space Intelligence Agency will build a spy station on or like the Moon.

Interestingly to note from a July 1970 survey that about 30% of Americans did not believe Armstrong and Aldrin ever walked on the Moon and it was just a NASA Hollywood stunt made movie on a soundstage. This was despite the huge amount of evidence, the dust and rock samples, television footage, 400,000 Americans in the NASA project and hundreds of thousands of scientists, engineers, and factory workers who made it happen. There are believers and nonbelievers.

SETI
On 20 November 1984 SETI Search for Extraterrestrial Intelligence was established with the mission to understand and explain the origin and nature of life in the cosmos and the evolution of intelligence.

SETI is a nonprofit science institute headquartered in Mountain View, California to conduct research and fieldwork in support of NASA and the National Science Foundation. SETI scientists are established all around the world for continuous tracking of promising signals from outer space.

Signatory signals coming from the vicinity of another planet or star are kept under continuous observation 24/7 with colleagues at their locations worldwide as the identity of the source sets and rises in multiple sites at the same time. The lookout for such widely differing phenomena as fast radio bursts, pulsars, and distinct artificial signals from extraterrestrial intelligence are under constant surveillance from roaming contact technology.

The same extraterrestrial signals coming from outer space require processing capabilities simultaneously at two locations using independent observatories. This is to verify and rule out that the signals were not caused by a technical glitch at any single site. The objective is to pinpoint the source and determine whether it has the characteristics of a signal from an extraterrestrial civilization.

Moon the #1 Defense Outpost

The key to the defense of Earth from the extraterrestrial is the Moon. The Moon is the most valuable prize in the extraterrestrial for Earth.

Were there to be a war among the countries on Earth or a war with the extraterrestrial, the Moon would be the determining factor as to the outcome and who is the victor.

The first line of defense for planet Earth in the extraterrestrial is the Moon for its strategic location and position. It is the Space Intelligence Agency base of surveillance outpost for Earth.

In preparation for the next major war, a couple of things need to be acknowledged. They deal with the admissions of we are not ready to fight the next war.

First ... it is important to admit we know nothing about fighting a war in the extraterrestrial.

Second ... it is vital to admit we do not have the necessary logistics in place to support refueling needs and the restocking of necessities, such as medical supplies and armaments for war.

What is vital to admit is what we do on Earth does not matter as much as what we do in the extraterrestrial. The extraterrestrial is more than our destiny, it is the key to our future and everything we know about our species.

Between what we think and know, imagine and experience, we cannot be prepared without making the unthinkable understandable and making the sacrifice of lives initially to explore with no return.

Experience comes from the past and preparedness comes from anticipation.

Until such time we reconcile our future with blind-activity, we will not make the necessary strides to prepare for the next events.

Rock – Satellite - Spaceship

More than a planet, more than a star, Earth is a laboratory.

There are hidden indications, extraterrestrials in finding a species for genetic adaptation, Earth might have become a genetic experiment. If we think of this as our origin or developed destiny, recent science and evolution of technology may conclude, we then should accept the existence of ETs and UFOs.

If so, we would furthermore determine the Moon is not a planet or a natural-made Moon but a spaceship with a spherical structure with a hollow-core as thought by some scientists. These scientists determined the Moon we see every night in the heavens was located there by design with an extraterrestrial intelligence behind it.

For a long-time scientist have theorized the Moon has a core that is hot and molten as thought the core of the Earth. No longer. On 19 November 1969, the Moon is hollow experiment was evidenced with the success of the Apollo 12 lunar module Intrepid mission. The experiment conducted crashed a satellite into the surface of the Moon. The result was a vibration of energy extending 20-miles with a sound level consistency for more than 1-hour.

The Moon was found to be a hollow-engineered artificial spacecraft with a metallic barrier solid surface. This scientific experiment concluded the Moon is a UFO spaceship flown in an orbit. Some infer this was just different geological and seismic forces at play. We won't know for sure until we go below the surface of the Moon.

July 1970 Moscow Russia announced at the renowned Russian Academy of Sciences the Moon is an artificial satellite put into orbit around Earth by intelligent beings of unknow origin.

Some of our best scientists are now in agreement the Moon is artificial in origin and raises questions we cannot answer and some we are not prepared to answer. Moon peculiarities find existing deposits of chromium, titanium, and zirconium which protect against extreme temperatures, radiation, and impact of mar meteorite bombardment. You just can't make this stuff up.

Tomorrow we will know more than we know today.

Understanding the Moon

It is tomorrow and we know more in visits to the Moon from evidence collected revealing huge amounts of data.
First President of NASA Commission of Lunar Exploration Astronomer Planetary Physicist Robert Jastrow

July 2020 German planetary geophysicists now believe the Moon is 85 million years younger than previously thought. It is believed the Moon was formed when the protoplanet Theia smashed into a nascent Earth some 140 million years after the birth of our solar system.

According to current calculations, the Moon was born 4.425 billion years ago. This time-frame matches closely with the previously determined formation of the core of the Earth. In other words, it appears the formation of the Moon occurred at the very end of the formation of the Earth.
Department of Planetary Physics at DLR German Aerospace Center Engineer Maxime Maurice
Institute of Experimental and Analytical Planetology at the University of Münster Prof. Dr. Thorsten Kleine

Over the centuries there have been many ideas put forward about how and when the Earth and Moon were formed. We really don't know a lot of what we are talking about. We haven't figured it all out.

There are those who theorize. There are scientists who look only at the physical evidence. There are others who spiritualize the creation of the Earth and Heaven.

As we are now aware, the physical scientific exploration of the Moon and our solar system is just beginning. We are starting to learn that many of our past theories, myths, and legends were wrong assumptions. The truth is we have very little information about the origin of the Moon and know very little about the purpose of the Moon. We still have a blindside and a lot to discover.

It may take a generation smarter than us to figure all this out. In the meantime let's keep all our options open.

Hollow-Moon

Throughout the history of Earth the Moon in the night sky has always fascinated. Yet, there is speculation the Moon is actually a terraformed and engineered piece of hardware with a 3-mile thick outer layer of dust and rocks.

Beneath the layer of dust and rocks is believed to exist a solid shell of around 20 miles of highly corrosive resistant materials such as Ti titanium 24, U uranium 236, Np neptunium 237, and brass generally of copper and zinc. These are rare elements that would not normally be expected to be found inside the Moon.

There are so many mysteries surrounding the Moon, on the other hand, causing some to believe the Moon is something entirely different. There are some who believe the Moon is actually a designated center of operations by an alien civilization for observation of Earth and the first strategic base for war in the extraterrestrial.

In 1970 the Soviet Academy of Sciences Scientists Mikhail Vasin and Alexander Scherbakov advanced the theory the Moon is hollow, an artificial satellite put into orbit around Earth, an ET alien intelligence base, and an extraterrestrial spaceship having a substantial interior space. The evidence to support this idea was based on 7 discoveries ...

#1 NASA 1969 experiment on the Moon indicated it seems engineered.

#2 The Moon has rare elements that should not be there.

#3 Scientists are nearly 100% sure the Moon doesn't have a solid core.

#4 The Moon is older than the Earth by nearly 800,000 years.

#5 The Moon is the only planet satellite that stays in a stationary orbit.

#6 Ten times more rich Titanium rocks on the Moon than Earth.

#7 The Moon maintains a precise altitude, course, and speed to Earth.

The Moon in the morning does not shine as bright as the Moon at night.

Space 10
Far Away Habitat Galaxies ... Not So Far Away

Reach the Future Quicker
Advancement in space travel and exploration should be for our benefit.

The outer space industry is set to be the largest employer and industry of this century. For the young it will be their nemesis.

The challenges and obstacles of the younger generations facing extraterrestrial competition for the better jobs in the outer space industry will be formidable. They will adapt in ways yet known.

When thinking about the future, it is important to see the long-term advantages in front of you. Don't find yourself mind-trapped by a deluge of noise and cloud of opinions from others. You would be better off to isolate yourself early on and determine how best to make your way and find your destiny in the space age.

A frontier with endless exploration and life experiences of a futuristic dimension awaits those brave enough to slip beyond the gravity of Earth. The sooner we reach for the stars the better.
The quicker the better.

What is happening silently and invisibly is where you want to be. The governments of the world know and are involved intimately.

Far Away Habitat Galaxies
Every existence of being has an image and a reality.

Images are created by opinions of governments sewn in the minds of their population. The reality is created by people who have traveled and lived in the dream. The truth lies somewhere in between.

Impossible you say. Non-human extraterrestrials do not exist since you have not personally met one face-to-face. To you, they are imaginations, fictional movie characters, and stars in television shows.

For most who have never seen something with their own eyes naturally doubt, and they should, that extraterrestrial life exists other than on planet Earth. Yet, there are those few who are brave enough to say they saw that, lived that, and know that.

All thoughts start from something and come from somewhere. You got yours somehow.

Okay then, where did these ideas come from ... ?

80 Species of ETs on Earth
Governments and their ministries of defense have often admitted they have identified at least a variety of 80 species of extraterrestrial aliens residing among us on planet Earth. There are more unidentified.
Canadian Defense Minister Paul Hellyer

Confessions of contact by humans on Earth, military intercepts, and government sanctioned meetings offer proof and documented evidence that extraterrestrial relationships exist worldwide with military and government officials. While confession is said to be good for the soul, it does not convince the majority of unbelievers.

We learn there are competing agendas. Not all are favorable to humans on Earth. Earth is not the protector but the protected.

Reportedly, two species have been identified who want to reduce the population of planet Earth. So it seems there are conflicting interests and wars over control of planet Earth between extraterrestrial civilizations from far away galaxies.

One group of astronomers and scientists have spent the last three years working on an image of the deepest region in space ever pictured to draw out unseen galaxies. The result is an incredible composite of various patterns unlike anything ever seen. The fascinating image was a combination of hundreds of pictures taken by the HST Hubble Space Telescope at NASA during 230 hours of observation in 2012. It showed a tiny region in the Fornax constellation thought dating back 13 billion years, a few hundred years beyond the conceived Big Bang theory like the Earth is flat.

The HUDF Hubble Ultra Deep Field image is a spectacular display of about 10,000 far-away galaxies. Some galaxies were discovered to be twice the size as originally thought. Astrophysicist Alejandro Borlaff at the IAC Institute of Astrophysics in the Canary Islands

Having considered an overwhelming preponderance of evidence, the existence of ETs and UFOs are out there. What we don't know is their agenda ... for us and planet Earth.

UFOs in Combat a National Security Threat
ETs flying UFOs from the extraterrestrial have always been with us.

UFOs have appeared in every war, cloaked in the day and at night to the human eye and radar. They are always around.

Looking back 160-years after the Civil War and every war since, militaries of all nations have recoded instances of sightings. War seems to increase frequency of visitations.

The first recorded instance of an enemy UFO attacking an earth aircraft was in 2004 in Italy. We are finally taking serious UFOs from the extraterrestrial as a national security threat. In the past we have experienced passive fly-bys. Now we have been aggressively attacked.

No military or Space Force on planet Earth has the technology to defend their nation or population. The best minds in the US military and intelligence know UFOs are flying reconnaissance missions searching for something, conducting mapping operations, scanning below the surface, and collecting intelligence throughout the world.

The Italy instance revealed that UFO spacecraft use a powerful electromagnetic ray-beam weapon. It operates in a select band of frequencies to minimize a retaliatory defense of an enemy.

We know an enemy UFO energy ray-beam weapon has the capability to take control of a missile and send it back to destroy its enemy who shot it.

This was a shot over the bow. The next one may result in loss of human life and mass destruction of property.

The conclusion was these UFOs are not US military craft. They are either foreign craft or more likely extraterrestrial spacecraft.

This makes for a national security risk that is real and should be taken seriously, investigated, and understood. It is never too late.

How to Defeat Enemy UFO Spacecraft

Technology is used every day and advanced weaponry is strictly only used for deterrence. All this may change.

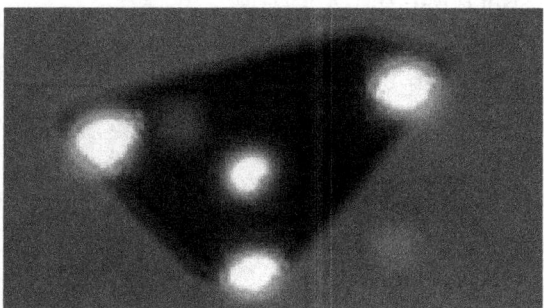

Anti-Gravity, Instantaneous Acceleration, Hypersonic Velocity
Low Observability, Trans-Medium Travel

This is an identified UFO Reconnaissance Spacecraft seen in many parts of the world. It has no sound and no radar identification. However, it makes a faint pulsing humming noise at a very low frequency and has three orange bright lights in the corners of the isosceles equilateral triangle with one red light in the middle underneath.

Search and Destroy

We now know how to track and destroy UFO spacecraft.

One way to track an UFO enemy spacecraft is to use the same frequency radio-beacon the UFO is using. UFOs can cloak and make itself invisible to military pilots and humans.

We cannot see a UFO in ambient light if the UFO is cloaked. UFOs can be seen in complete darkness using infrared night-vision which measures temperature differences in varying signatures.

An UFO is vulnerable to attack when it is outside its stealth mode. It cannot use its energy weapons when cloaked. Then we can attack with uranium grade weapons while the UFO is not shielded.

American astrophysicist and cosmologist Carl Sagan believed that UFO claims required extraordinary proof. Secretary of Defense for Intelligence Chris Mellon believes we now have the proof.

So What Should We Do?
Maybe find consolation in elevating our education programs to more meaningful subject matter which can support the knowledge-skills needed for the space age of tomorrow.

Maybe find comfort in knowing we have never come face-to-face with an ET or UFO.

Maybe find solitude in the thought we are not alone and maybe entering a new plane of existence.

Maybe find satisfaction in realizing there is more to life than our small planet in the large scheme of phenomena.

Maybe find solace in the confidence of a Space Intelligence Agency out there to protect us from whatever we next confront in the extraterrestrial.

This has become a defining moment in our existence on planet Earth with an existing alien presence we are just beginning to learn how to deal with for our benefit.

Educate Future Generations
As an educator for many years, I can say without hesitation our academic curriculums and programs are obsolete for the space age.

To tackle the space age challenges, it would help to have an advanced space-age curriculum, both academic and vocational, that would prepare our youth for the jobs in the extraterrestrial.

Preparation precedes participation. Space-age preparatory subjects would include foundation courses in astrophysics, astronomy, engineering, physics, chemistry, biology, geography, space sciences, computer science, mathematics, medicine, chiropractic, psychiatry, linguistics, agriculture, anthropology, communication, and business. There are more which don't exist at present.

As older generations pass, younger generations will find their way to the stars.

Explore Outer Space

The wanting to know to answer questions of curiosity and the ambition to be first are good reasons for the NASA and space forces to explore outer space.

Another main ambition to explore outer space is profit. This would bring the cosmos within Earth's sphere of economic influence.

Being first is in our genes. Making money is how we occupy our lifetime.

The idea of learning is part of our curiosity-makeup. Discovery of things and new knowledge fulfills our lives with satisfaction.

As Earthlings, we strive to accomplish things for reasons that are intuitive and compelling. We do things not because they are easy or logical.

We do things because we can and they are emotionally compelling. We want to know how to find immortality.

Many want recognition. Some want to be heroes. A few want anonymity.

Every civilization builds monuments to remind us they were here. It seems it is a nature solely not confined to Earth but to other planets and civilizations in faraway galaxies.

We explore the universe to find ourselves and our origins. We seek answers to questions not yet asked.

We want to know who our extraterrestrial neighbors are. We want to know if they are our friends or a threat.

We want to make a better world or worlds and a better life for our species. Our compassion to explore is infinite.

The reason we explore outer space is because it is there.

Final Thoughts of the Author

The idea for this book began with a vision for a space intelligence agency some thirty years ago.

Today the space age has launched an interest in space travel for humans to the stars and planets, in mining for rare and raw materials on other planets for business and exploration of the cosmos for pioneering settlements in other worlds.

There are a thousand reasons why many people don't believe in the existence of ETs and UFOs. The good news is there are a million reasons, all surmountable, why ETs and UFOs live in our world.

We were living in an unnatural state of denial when I started compiling a list of the best of the best knowledge about ETs and UFOs. Nobody could have predicted where we would be in our understanding of outer space today.

The list of made-known and publicized ET and UFO events, sightings, and experiences were exhaustive and when combined were a drop-in-the-bucket as compared to what is out there.

I discovered in my research it drastically omits a vast listing so comprehensive as to capture what has actually occurred from the extraterrestrial on planet Earth. And yet, we are here still testing the waters of believability.

Over the past few years, the Earth has become uncertain and traumatic. Yet, there have been huge sacrifices and hardships by those few brave coming forward with their stories about ETs and UFOs.

They have shone great examples of the human spirit to what is known to be the truth. We are living in the most extraordinary of times.

The revelations of ETs and UFOs have obliterated what we thought was normal. The very idea of a natural-type-world rightly has been redrawn and recast into a definition many can agree on while most are still unwilling to accept. This brings empathy to the conscience, seeking to understand rather than impose.

With unconditional acceptance of extraterrestrial phenomena with all its trappings our world has brought about a shift in social change. The past decorum of tradition of the very idea we aren't the sole occupiers of the cosmos seems absurd to most people.

Learning and developing the skills needed for the coming outer space age offers value for the sale of goods and services to drive your life and business forward. All life on planet Earth is business in one form or another.

To accept the premise the highest standard of living for the future will be discovered working in the extraterrestrial is to admit life exists beyond Earth. This is the prevailing wisdom I leave with you. Take seriously the long-term view, the inner of Earth and outer of space will fuse us all into one new reality.

The new way of living with phenomena inspired by timeless highlights and silhouettes awakenings is to revisit the past. Our newest outposts may take us to the Moon and beyond.

There have always been a lot of great ideas in science and technology, but never the money to do the expanded research, unconventional experiments and scientific testing until now.

Where science and technology leave off, imagination propels us into the future. Now there is a vision of life for the times to come.

Traveling in outer space will bring together long-held conventions of a whole new world. As we glide among the stars and navigate the planets, we will always be in good company, in an *Alien Presence*.

ACKNOWLEDGMENTS

Great books are never produced alone.

Renee' Rogers Grooms
Amanuensis
The University of Texas at Tyler M.Ed.
Texas A&M University at Commerce B.S.

First and foremost, I would like to thank my wife for her participation. She has been incredibly supportive throughout this process. Thanks for always being there.
Writing a book requires a lot of time in isolation.

Christopher Scott Puckett
Editor and Proofreader
Southern Methodist University M.B.A.
The University of Texas at Austin B.B.A.

My gratitude to my son-in-law for all the time from your family to improve this writing. Your keen eye for detail and challenges have made this a better book. Thank you for holding me accountable and for your felicitous ways.

Dr. Jacqueline Charlene Grooms
Graphics and Book Cover Designer
Parker University D.C. B.Sc. – M.B.A. – B.B.A. VAL
The Hockaday School

Finally, a special thanks to my daughter for giving of her creative talent. To put the polish on the finishing pieces, I depended on the advice, conversations, and sustenance of your contribution.

BOOKS

To be free it is necessary to free our mind. Read to have a foundation of understanding. The greatest successes are the individuals who spend almost all their time reading. New ideas are born from cumulative knowledge over time. Ideas are a consequence of vociferous reading. A reader can enjoy several lifetimes in one.

ABOUT THE AUTHOR

Thomas Fletcher Grooms is the #1 bestselling author of nonfiction books on outer space intelligence and business in the extraterrestrial.

He is the recipient of numerous awards including the Texas Authors Nonfiction Historical Award.

Tom lives in Denison Texas with his wife. You can learn more about Tom by visiting his website at https://tomgrooms.com.

The Market Intelligence Collection

The Outer Space Intelligence Series

The Russia Series

The Executive Leadership Collection

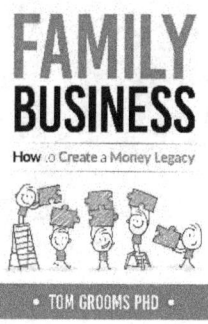

CPSIA information can be obtained
at www.ICGtesting.com
Printed in the USA
BVHW040208280421
606012BV00012B/345